Niche Envy

Niche Envy

Marketing Discrimination in the Digital Age

Joseph Turow

The MIT Press
Cambridge, Massachusetts
London, England

For information on quantity discounts, please email special_sales@mitpress.mit.edu.

Set in Stone Sans and Stone Serif by the MIT Press. Printed and bound in the United States of America.

Library of Congress Cataloging-in-Publication Data

Turow, Joseph
Niche envy : marketing discrimination in the digital age / Joseph Turow
   p. cm.
Includes bibliographical references and index.
ISBN-10 0-262-20165-8 (alk. paper)
ISBN-13 978-0-262-20165-0 (alk. paper)
1. Consumer profiling. 2. Market segmentation. 3. Marketing—Technological innovations. 4. Customer services—Technological innovations. I. Title.

HF5415.32.T95 2002
658.8'34—dc22

2005058409

10 9 8 7 6 5 4 3 2

Contents

Acknowledgements

This project could not have been carried out without the help of tens of individuals from media companies, marketing communication firms, banks, supermarkets, department stores, and consulting firms who patiently let me question them about their work. I mention many of these people by name in the book. I don't cite others (sometimes at their request), but their observations were nevertheless crucial in pointing me in important research directions. I hope that marketing and media industry personnel reading these lines will share some of their valuable time with academic researchers who contact them. Apart from being kind, they will be helping to contribute a multiplicity of viewpoints to writings that may become central to public discussions of their business.

In many cases, I contacted sources because they were mentioned in articles I read in trade magazines or because they appeared on panels at industry meetings I attended. The industry meetings were useful for confirming, extending, or refuting what I learned in "the trades." I have been reading *Advertising Age* and *Variety* habitually for about four decades, and I have been a visitor to Mediapost's online daily newsletters since their inception several years ago. This literature has helped to shape my understanding of the basic contours of the media-and-marketing system. In addition, the Nexis and Factiva databases have made it practical to explore particular topics in depth across a wide range of industry periodicals (such as *Adweek*, *PR Week*, *Progressive Grocer*, and *Chain Store Age*) and government agency publications. I have also benefited from reports from media and marketing consulting firms, such as Forrester Research, Jupiter Research, and E-Marketer. I am grateful to the University of Pennsylvania's Annenberg School for Communication and to the university's library system for providing access from my office desk to these and many other rich intellectual resources.

Michael Delli Carpini, dean of the Annenberg School for Communication, has provided an encouraging environment for research through his personal style and his adjustment of the faculty course load with an eye toward the production of knowledge. Some years ago, Kathleen Hall Jamieson, Director of the University of Pennsylvania's Annenberg Public Policy Center, asked me to head the Center's Information and Society division. The post came with the funds to carry out the national surveys that I discuss in chapters 7 and 8. All this has taken place within a school whose faculty members not only have been collegial but have really cared about and supported one another. It has been a great environment in which to work, and I thank my colleagues for fostering it.

Annenberg School graduate students Mary Bock, Lee Shaker, and Lokman Tsui have tracked down documents for me and, more important, have critiqued portions of the evolving manuscript. My thinking on topics in this book has also benefited enormously from discussions with Jatin Atre, Beau Brendler, Rich Cardona, Kyle Cassidy, Eszter Hargittai, Chris Hoofnagle, Elihu Katz, Toby Levin, Rob Mayer, Matt McAllister, Lee Rainie, Marc Rotenberg, and Jonathan Turow; I apologize to anyone I have accidentally omitted. At The MIT Press, Robert Prior has been an enthusiastic and supportive editor, and Valerie Geary has been a congenial contact person.

I have particularly benefited from discussing evolving book ideas with my wife, Judy Turow. She has had the patience to hear me out, be supportive, and critique me honestly, even in contexts (movie theaters, family trips) where my raising these rather specialized topics probably was less than appropriate. I thank her most of all.

Niche Envy

1 :: A Major Transformation

We argue strenuously, strenuously against the naive sentimentalism on the part of companies that insist "We love all our customers and we love all our customers the same."
—advertising executive quoted in *Advertising Age*, March 1995

[These customers] don't spend much money with you now, aren't big spenders in the category with your competitors and, for whatever reason, lack the capacity to increase consumption in your category in the future. . . . If you can avoid recruiting them into your program from the beginning, do so. In many cases, however, until they have joined the program, you have no way of assessing their value. . . . The goal is to starve them out of the program quietly but effectively.
—loyalty consultant Richard Barlow, October 2000[1]

When they were written, those comments were meant to be provocative, even controversial. Today, however, the reasoning they represent is conventional among marketers. At their most politically correct, they speak of a "customer-centric approach." In the words of one writer, "all employees of a company, from the CEO on down, must continually ask themselves what would they like if they were a customer of their company."[2] But as the two quotations above suggest, cold winds of change are pushing executives toward tough decisions as to which customers really count and how to talk to them as personally and as customer-centrically as is practicable. Marketers increasingly use computer technologies to generate ever-more-carefully defined consumer categories—or niches—that tag consumers as desirable or undesirable for their business. Increasingly, too, they use computer technologies to vary the content and the scheduling of messages they send to people in different niches.

This book is about how the movement of databases to the heart of marketing communication is beginning to affect the media, advertising, and society. Media and marketing practitioners recognize that their businesses will come to be centered on data-driven relationships with customers whom they care about and who care about them. In the twentieth century, Americans came to think that just about all members of their society had access to certain kinds of knowledge about products—their existence, their ingredients, their range of prices, how they might be used, and what public images their sellers wanted to associate with them. Now, however, there is an impetus to make that common access disappear. Marketers' new goal is to customize commercial announcements so that different people learn different things about products depending on what the marketers conclude about their personalities, their lifestyles, and their spending histories. To help marketers do that, media firms will increasingly deliver different advertisements, different programs, and even different parts of programs.

At first glance, the idea of such customizing may not seem at all objectionable. It might, in fact, seem to benefit individual consumers. Some advertisers will give certain consumers great discounts. Some media firms will vary news and entertainment programs to match what certain consumers like, so they will tune in to the ads, which themselves will be appropriately personalized. Optimistic executives insist that such customization will "satisfy [the] difference and diversity" of the American population.[3]

Yet this book's tour of the industrial logic behind the activities makes clear that the emerging marketplace will be far more an inciter of angst over social difference than a celebration of the American "salad bowl." Advertisers want consumers to worry that they will not get desirable discounts and media materials unless they offer up information that will help the advertisers to customize persuasive messages specifically for them. Advertisers also want customers to know that to be favored with the best deals they must reveal attributes and activities that make them especially valuable to particular advertisers.

At times, individuals may be happy to get what they want when they want it. Over the long haul, however, this intersection of large selling organizations and new surveillance technologies seems sure to encourage a particularly corrosive form of personal and social tension. Audiences will quite logically assume—in fact, they may even be told—that the customized advertisements, entertainment, news, and information they receive reflect their standing in society. They may be alarmed if they feel that certain marketers have mistaken their income bracket, their race, their gender, or their political views. They may ask themselves if the media content that friends or family members receive is better, more interesting, or more lucrative than theirs, and whether the others have lied about themselves to get better material. They may try to improve their profiles with advertisers and media firms by establishing buying patterns and lifestyle tracks that make them look good—or by taking actions to make others look bad. Such responses to the new importance of niches should not be considered social paranoia. They will flow directly out of the developing logic and structure of database marketing. There already is resistance to these developments, and there may well be more. Yet the competitive factors shaping database marketing and the media technologies connected to it are so strong that social criticisms will not derail them. Instead, the public rhetoric about data-driven personalization in marketing will likely be ever more rosy. Marketing and media executives are already proclaiming that it will increase attention to particular customers and therefore reduce their chances of experiencing bad service and identity theft. I will show, however, that, by emphasizing the individual to an extreme, the new niche-making forces are encouraging values that diminish the sense of belonging that is necessary to a healthy civic life.

::

"Niche envy" has two meanings. One meaning pertains to competitors, who may envy the "quality" of other competitors' customers. The other pertains to consumers, who may envy what they believe are their friends' better profiles, which may get them better treatment from media companies, from stores, or even from manufacturers. Both meanings suggest that somehow the marketplace is deeply involved in defining an important basis for belonging in society.

It may seem strange to associate the marketplace with a sense of belonging. Yet it has long been true that the marketplace is more than an arena in which people can buy stuff; it is the hub of social life. The complex, industrialized American marketplace is no exception. It is hard to think of any part of life that is not continually affected by it. The Christmas season stands out as an example. Over a century and a half, it has been fashioned by the market into a period of near-frantic consumption. The holiday's approach signals the year's most important selling season. As the news media remind Americans every year, economists see Christmas shopping as a barometer of their society's fiscal health. Yet the social influence of the modern retail system also permeates the most mundane activities: trips to buy food, journeys to shopping malls, visits to websites such as Amazon and Expedia. Many of these experiences are pleasant, though not all of them and not for everyone. Some individuals feel as if they will break out in hives even at the thought of going to a mall. Perhaps these same people cringe every morning as they delete email blandishing products from loans to erotica. They may mutter angrily when, upon coming home, they carry third-class postal mail directly to a wastebasket. More serious damage—both emotional and material—is caused by market fraud and defective products. Blatant negligence and irresponsibility are other areas of concern. For example, in 2004 critics accused America Online of sloppy oversight of customer records after an employee stole its database of 30 million subscribers and sold it to "spammers" for $100,000. While AOL denied that its information-handling procedures were sloppy, observers predicted class-action lawsuits by clients who felt abused.[4]

Over centuries, intricate mechanisms have been put in place to regulate relationships between sellers and buyers. Large bodies of law on fraud, negligence, and restraint of trade set the basic boundaries of corporate activities and consumer resistance. Complementing them are rulings by government commissions that aim to keep the marketplace socially acceptable and predictable. All these matters are open to negotiation and struggle. Companies and industry groups hire expensive lobbyists in order to influence what the law says is acceptable. They also try to persuade government officials of the benefits of letting the industry regulate itself. Nonprofit advocacy groups, less well funded, try to ensure that new developments in technology, competition, or corporate policy do not allow companies to make "end runs" around what has become a social norm: the public's right to a competitive, honest marketplace that is available to all comers, with open access to information about products.

The norm has not quite been matched by the facts. That honesty in the sales arena still needs policing is an idea widely shared through jokes about used-car dealers, web warnings about identity "phishing,"[5] and news reports about price fixing by major merchants. The openness of the marketplace gets much less attention than its probity. As with honesty, the record is mixed. Marketers often succeed in hiding information about pricing strategies that might affect profit margins. A pants manufacturer who sells the same slacks through J. C. Penney and Brooks Brothers under different names is unlikely to advertise that fact and indeed might prohibit J. C. Penney from disclosing it. Then, too, certain classes of people have traditionally been favored with advice about the availability of products aimed at them and their lifestyles—direct marketing to the very wealthy by means of personal visits, letters, and special DVDs is an example.[6]

At the same time, a strong case can be made that it was fortunate for the twentieth-century consumer that public access to information about goods and services became relatively open—one might even say relatively "democratic"—even as industrial power took greater control of shopping. For historical reasons that I will discuss in chapter 2, most Americans today find it possible to learn about the existence, the nature, and the pricing of all sorts of products, even those not intended for them. Government regulation has been of some importance here; think of the ingredient-labeling requirements for foods. The two most important democratizing vehicles for consumers' awareness of products, though, were the large department store and advertiser-supported media.

It is hard today to imagine the enormous change that department stores brought in people's awareness of the world of trade when they replaced itinerant peddlers and small merchants. At the most basic level, these predecessors of today's shopping emporia allowed people from a wide range of circumstances to see an enormous display of merchandise offered at publicly posted prices. Advertising-supported mass media reinforced this accessibility. Anyone who chose the same magazines, newspapers, radio stations, and television channels as anyone else would see or hear the same commercial messages, many of them announcing sales. Moving across advertisements and stores, steady customers, would-be customers, and cultural voyeurs were able to learn about various products' components, their uses, and the range of prices at which they would be available. These features and others would also let consumers in on the social meaning that the company wanted to attach to the product—for example, whether certain clothing was intended for casual, business, or evening wear, whether for the frugal or the indulgent shopper, whether for the trendy or the traditional, or whether for the full-figured, petite, older, or younger woman.

Many consumers' encounters with advertising were elements of a symbiotic relationship that emerged in the twentieth century among consumers, marketers, and media firms. Marketers were able to present products and their "personalities" to huge audiences. Media firms received money from marketers to help them create materials that would attract people to advertisements. Consumers received entertainment, news, and information free or at costs far below what they would have to pay for commercial-free media. As the historian Daniel Boorstin notes, in the process they were also inculcated with the idea that they were linked to millions of other people who were also sharing these same products and lifestyles.[7] One reason this worked was that marketers felt they had the better part of the bargain—and indeed they did, as I will explain in chapter 2. An equilibrium developed among advertisers, media, and audiences that allowed consumers access to a remarkably predictable marketplace culture. Underlying that equilibrium, and infusing that culture, was a mindset that became known as "keeping up with the Joneses." Marketers told consumers that feeling an almost painful envy of what others had was the American Way. Phrased most positively, "keeping up with the Joneses" meant "Large numbers of Americans live a great life by getting the latest products that everybody wants—and you should too."[8] Over the course of the twentieth century, "keeping up with the Joneses" became deeply enmeshed in American society.

Now the equilibrium has begun to falter. Marketers know that American consumers are using "pop-up killers" and digital video recorders to avoid advertisements in sponsored media. Simultaneously, consumers are using the internet[9] and other new media to learn more than marketers want them to know about products, prices, and alternatives. Targeting of audience segments emerged in the 1970s as a way for advertisers to cope with the media fragmentation. Targeting has not, however, stopped audiences from trying to evade commercials. Marketers have come to believe that discouraging that activity will require more than trying to derail ad-zapping equipment. It will require building up customers' loyalty to particular firms so that they will look forward to seeing and hearing about those firms' products and services. With that in mind, marketers and their media allies are developing technologies that will enable them to go beyond targeting. Invoking the nineteenth-century economist Vilfredo Pareto, they have developed a new mantra: "Focus 80 percent of your efforts on the

20 percent of customers who provide 80 percent of your profit."[10] That dictum took on great interest in the 1990s under the rubric of customer-relationship management (CRM).

Customer-relationship management is based on the idea of cultivating "best customers" through direct mail, telemarketing, in-store selling strategies, and loyalty programs. For example, Bloomingdale's, a chain of department stores, uses a CRM system called Klondike to feed data about its 15,000 "most valuable" patrons straight to the chain's call center or to the sales floor of one of its stores. Klondike's database brings together records of transactions, promotion histories, basic household information, and even photos of customers. Live links to point-of-sale terminals help salespersons to offer customized services and suggest merchandise. According to the marketing magazine *Direct*, "when the store hosts one of its 'girls' night out' specials, a sales rep can be alerted that a given customer is particularly desired at the event, and can be fed information about it. That information is printed on the customer's receipt, too."[11] This is crucially different from the twentieth-century notion of consumer envy. It suggests that marketers have much more direct power over consumers than they once had, and that now consumers must work with marketers—even work to attract them. It suggests that people increasingly identify with niches rather than with the broad American middle or upper-middle class—and that they ought to do so. It also suggests that the best way to gain access to the good life is to release information that establishes one's desirability as a customer.

Sophisticated population databases are central to the new approach. Marketers increasingly examine their own files and files they buy or rent to determine whether current or potential consumers fit their requirements for desirable customers. If consumers don't fit those requirements, marketers can move in one of two directions: They can push the consumers away physically and electronically, or they can encourage them to provide data that attest to their value. To be favored with good deals and products in the new marketing world, a customer not only must allow surveillance but also must show evidence of his or her value. The financial-services firm Morgan Stanley said as much in a 2004 letter encouraging its "most valued clients" to reveal the separate assets they and relatives have deposited with the firm by collecting them into a "household relationship." The reward for doing this would be special benefits, the letter said.

The flip side would be punishment: "The Household will be reviewed annually to determine if it still continues to qualify for tiered benefits. Households that do not maintain the required balances for benefits programs may be downgraded or removed from those programs." The letter did not specify what those requirements are, which left open the possibility that it might vary by person. This example is merely the tip of a huge iceberg of emerging customer-marketer interaction. Marketers' increasing desire and ability to discriminate among individual customers on the basis of their contributions to the bottom line has become a badge of honor, an indication that they are following through on the potential of the digital age. Encouraging themselves and the public with such terms as "relationship," "permission," and "loyalty," marketers are already beginning to tailor advertisements and offers in ways that speak to each customer's unique combination of income, sex, age, geography, and lifestyle (and, more cautiously, race). They are beginning to link profile-driven content, product suggestions, and prices to various forms of interactive media at home, at work, outdoors, and in stores. The aim is to surround attractive customers with personalized commercial blandishments at times when they are optimally effective and in ways that can't be ignored.

National survey research conducted by the Annenberg Public Policy Center indicates that most American adults have a sense that niche-building activities are taking place, and that the information they offer up to websites and stores has consequences. They admit that this makes them nervous, and they overwhelmingly disagree with the statement "What companies know about me won't hurt me." At the same time, the Annenberg surveys show, American shoppers don't have a clue about how marketers use data on them, or even that such niche-building activities as differential pricing are legal. As marketers' new approaches to customers get more coverage in the news media, and as more and more people learn about them, a number of questions that in the past were hardly asked become more relevant:

• What ethical and practical issues arise when advertising and editorial matter are tailored to individuals on the basis of media firms' necessarily incomplete ideas about them?

• What is the social significance of executives' newfound insistence that consumers are morally obligated to pay attention to advertisements in return for the free or discounted media material they receive?

- In a society concerned with getting the best deal as well as with keeping personal information private, how should public policies respond to social divisions that are bound to grow as people envy the data files that enable their peers to get seemingly better prices, seemingly better service, or both?

One reason there is little social debate about such questions is that the concerns they reflect have become realistic only during the past few years, as database technologies have gotten more sophisticated. Just as important, though, is that until recently the only people paying much attention to database marketing were executives carrying it out and privacy advocates worrying about it. Led by groups such as the Electronic Privacy Information Center and writers such as Robert Harrow, Daniel Solove, Peter Suber, Simson Garfinkel, and Oscar Gandy, privacy advocates have drawn attention to ethical and practical issues related to the accumulation and the use of personal information about customers.[12] However, the use of databases in marketing communication also raises other issues. Critical observers of commercialism have only begun to get a handle on the quickly evolving scene of advertising, direct marketing, product placement, in-store displays and salesmanship, public relations, catalogs, and online selling. One reason they have hardly discussed the use of databases is that they have not developed a vocabulary for doing so, or even a map of the scene. For much of the twentieth century, advertising was the part that stood for the whole of marketing communication in popular writing and even in academic writing. This was influenced by top advertising executives' efforts to position their work as historically mainstream. In the nineteenth century, advertisements were certainly a common feature of the American landscape. Also common, however, were the sneaky blandishments of the carnival barker, the news-controlling shenanigans of the publicist, and the insistent claims of the door-to-door salesman. As they sought respectability in the early twentieth century, advertising practitioners didn't want to be associated with those activities.

Of course, the advertising business wasn't as nearly pure as Ivory Soap. Around 1900, the curative claims of ads for patent medicines were prominently questioned. Major publishers began to refuse to accept such ads, and the 1905 Pure Food and Drug Act was intended to get rid of the most dangerous patent medicines (which routinely contained alcohol, cocaine, and even arsenic).[13] Despite some amelioration, advertising in general retained its reputation for playing loose with the truth. Advertising executives knew it, but they also knew that they had important sources of goodwill among consumers. For one thing, advertising's support made newspapers and magazines inexpensive. For another, advertisements in periodicals and on billboards were a window on the consumer cornucopia that the industrial revolution had created.

Increasingly elaborate and beautiful advertisements, sometimes in color, depicted a broad menu of consumer products. By the 1920s, overt selling through print advertising (and radio advertising) had become taken for granted as part of American life. Particular advertisements—for example, those for Cadillac and Jordan automobiles, Listerine mouthwash, and Woodbury soap—drew the public's attention to those who created advertisements. The advertising business strove to act modern and to introduce Americans to modernity.

At the self-promoting apex of the product-selling hierarchy were large "full-service" advertising operations such as J. Walter Thompson. They embodied the three basic functions of advertising work: creative persuasion, media buying, and market research. The "creatives" of the agency—copywriters and art directors—created print, radio, and (in the late 1940s) television advertisements based on input from market researchers and clients. Media buyers then purchased space or time for the ads.

By the late 1950s, the relationship between large advertising agencies and media firms had become routine. Because of the huge amounts of money their major clients spent on advertising, the ad agencies wielded a great deal of power over the media. Magazine publishers and television networks understood that they were obligated to deliver large national audiences who could see the ads that the agencies placed for their clients. Of particular importance were the three commercial television networks (ABC, CBS, and NBC) and the large magazine firms, including Time Incorporated, Triangle Publications, and Meredith. Though pressures for creativity, audience reach, and sales created tensions among practitioners, there was a strong predictability to the relationship between advertisers and the media.

Popular commentators accepted the advertising elite's picture of the persuasion industry. In the 1958 book *Madison Avenue, USA*, Martin Mayer defined his subject as "national advertising of branded products in general media of information and entertainment."[14] In *The Story of Advertising*, also published in 1958, James Playsted Wood framed Mayer's tale historically, concluding that advertising as he defined it had become the central aspect of commercialism and was "ubiquitous, incessant, and inescapable."[15]

Reacting to the print, radio, and television commercials that surrounded them, academics in the fields of sociology, history, cultural studies, literature, and communication explored the same terrain that the advertising community and popular writers investigated, but with a colder eye. In the ensuing decades, their analyses drew a fascinating map of mainstream advertising's work as a cultural and ideological arm of business. But they often wore blinders when it came to what advertising people called "below-the-line" marketing communication, even though it went way back to the early years of the advertising industry. A pre-World War II example is Edward Bernays's public-relations work that complemented the Lord and Thomas ad agency's campaign encouraging women to smoke Lucky Strike cigarettes.[16] Another is Ovaltine's giveaway of Little Orphan Annie mugs to listeners of the *Annie* radio program who mailed in Ovaltine labels. Such activities weren't central to advertising histories.

In the 1990s, though, it became painfully obvious to many in advertising and academia that the center had shifted dramatically. The kinds of activities from which the leaders of mainstream advertising and media had tried to distance their work in earlier decades now flowed out of the biggest firms in such forms as telemarketing, guerilla marketing, stealth marketing, viral marketing, in-store displays, email, sampling, and free-standing inserts. Sales tactics were integrated with public relations in the creation of entertainment, information, and even news.

Some writers recognized the new centrality and interdependence of such activities. In 1996, Matthew McAllister noted in *The Commercialization of American Culture* that the time had come to talk about "new forms of advertising" that were influencing both new and traditional media.[17] Four years later, Thomas Frank, in *One Market under God*, discussed public relations, e-commerce, product placement, and event marketing in addition to conventional overt advertising in his critique of "extreme capitalism."[18] The observers most likely to confront the changes taking place were academics or activists with specific social concerns. For example, Jeffrey Chester, head of the Center for Digital Democracy, spoke out about what he considered the anti-democratic features of privileged media domains ("walled gardens").[19] Alex Molnar of Arizona State University's Commercial Education Research Unit took aim at product placement in schools.[20]

Yet a coherent map of the world of marketing communication is still to be drawn. The perspectives of critics often do not overlap. They also do not fit into a coherent explanation of the evolving relationship between traditional advertising and other forms of marketing. Looking to advertising people for enlightenment is of only limited help. Though knowledgeable about particulars, they are often hard pressed to come up with nuanced explanations of the changes that are occurring in their business and the reasons for them. However, advertising people understand that the very structure of their industry is changing drastically, almost traumatically. The major unit of the "advertising" industry is now not an agency but a holding company, such as U.S.-based Omnicom or Interpublic, Paris-based Publicis, or London-based WPP. Each of these companies has under its umbrella a dizzying number and array of firms, including ad agencies, media-buying operations, public-relations firms, branding and identity firms, research consultancies, and firms that engage in such previously marginalized marketing communication activities as direct marketing and promotion. Like Russian nested dolls, some of the conglomerates' firms— for example, WPP's venerable J. Walter Thompson agency—are themselves large conglomerates. These conglomerates do not privilege advertising agencies as the prime movers in marketing communication; sometimes, a branding agency or a "buzz marketing" subsidiary takes the lead in guiding a client. In 2005, when the heads of WPP, Omnicom, and Interpublic acknowledged this development, it made the front page of the trade magazine *Advertising Age*.[21]

The changes on the media side are just as rattling. The copywriters, art directors, researchers, media planners, and media buyers of the national advertising establishment no longer focus on radio stations, magazine, and three television networks. They now focus on huge media conglomerates (Viacom, CBS, Disney, Sony, Time Warner, News Corporation, Clear Channel), which have under their wings movie firms, magazines, radio stations, television networks, and concert venues that seem willing to work with agencies on marketing communications activities that they would have shunned not long ago. Beyond the conglomerates are fragmented in-home media worlds (with VCRs, DVDs, and telephones) and fragmented out-of-home venues (from supermarkets to doctors' offices) that are natural targets for all sorts of marketing communication. And if these new developments were not enough to place huge pressures on advertising practitioners to bring coherence and predictable equilibrium to the changes, the rise of the World Wide Web, with its spam and its interactive advertisements, began to bring whatever was left of the traditional model of the mainstream advertising industry crashing down.

In 2004, Scott Donaton, the editor of *Advertising Age*, told his readers that it would be necessary to "rewrite the definition of the word advertising" as part of a larger effort by marketers to "confront and release their historical biases" and "redefine their world."[22] Donaton did not say what that new definition would be, but he was clearly tapping into a consensus that new realities with huge implications for the business of selling were emerging.

Many in the business press have described the situation, or parts of it, in ways that reflect Donaton's suggestion of a sharp break with what went before. According to one *Advertising Age* columnist, "the business of advertising is under extraordinary pressure," and fundamental "cracks in the foundation" of audience research may help bring the entire house down unless fundamental changes are made.[23] An article in *Fortune* titled "Nightmare on Madison Avenue" claims that "the best way for Madison Avenue to begin is to let go of the past."[24] The vice president of a major research firm describes the expansion of ad-skipping consumer technologies as transforming the habits of mainstream consumers.[25] Bringing many of these concerns together, a writer for the British trade journal *Marketing* writes: "Earthquakes are sudden and violent. But long before they happen, there are shifts observable in the tectonic plates beneath the earth's surface. So perhaps it is time to get ready for marketing's own earthquake."[26]

::

It is important, in trying to understand what is taking place in the advertising system and what its implications are for the media and for society, to take views expressed in the business press seriously. But is a metaphor of fundamental break—an earthquake, an explosion, a letting go of the past— really the best way to think about what has happened and where things are going? Writings on the ways organizations and industries face complex new challenges point to a very different approach. They suggest that the best way to understand how advertising executives are planning the industry's future is to see how they are rethinking their industry's past. As I will show in chapters 2 and 3, features of marketing communication that many speak of as new—product placement and certain aspects of direct selling, for example—have been around for many decades. Today's media and marketing communication executives are re-imagining them as they confront new challenges.

One fundamental challenge they face has to do with the trust of consumers. Trust, Francis Fukuyama notes, is belief that an actor is involved in "regular, honest, and cooperative behavior, based on commonly shared norms."[27] Anthony Giddens points out that a trusting public is a critical resource for sustaining the organizations that make up institutions such as the government or medical system.[28] To keep their authority, these organizations must continually convince the public of their competence, integrity, and benevolence. Giddens might have added marketers and the media to his examples. It is an article of faith in marketing that customers or prospective customers believe that a product or a service should work as advertised. But practitioners of marketing communication insist that consumers' perceptions of integrity and predictability are also critical. A "personality" that stands for trust in the minds of relevant audiences is, they say, what ultimately constitutes a successful brand.

Often a brand's personality is depicted through stories. In the case of a magazine advertisement that details the joys of a Ford Mustang or a television commercial that extols Pond's cold cream, the ad creator's goal is to suggest a story that bespeaks the product's usefulness for the audience. Of course, to do that, the advertising people must have thoughts about the audience, particularly as it relates to the product they are selling. Using research data to help them imagine the audience's characteristics, a creative team can concoct a sales pitch. Moreover, marketers see the choice of a media environment as crucial to the cultivation of consumers' trust. The goal is to present a product in a media environment (a specific magazine, network TV series, or cable network) that the intended audience believes will deliver the news, entertainment, information, or education they expect. Marketers expect that this media trust will make it easier to get the audience to accept their messages. Despite their power over the media system, advertisers have long been dependent on media producers to deliver their ads to the right number and the right kind of people. Advertisers have also depended on media firms to lead audience members to believe that the presence of commercial messages is legitimate and to be taken for granted as a fact of life.

Marketers recognize that an important reason consumers look for companies and brands they trust is to avoid risk. At its most basic, risk means the possibility of buying shoddy merchandise or wasting time with bad media products (for example, a boring movie or an unexciting video game). The digital media environment has brought new concerns about consumers' unease. One risk of going online is that you may be bothered by advertising that you don't want. Another is that you may be giving companies personal information that you wouldn't want them to get.

As Oscar Renn and his colleagues have noted, risk can be conceived as both "a potential for harm" and the "social construction of worry."[29] How people understand a potential danger plays a large part in determining that phenomenon's importance as a topic in their society.

Today many marketing and media firms are struggling with strategic management of the public's perception of risk and trust. Their goal is to persuade desirable customers to trust them with personal information on the understanding that it will bring those customers benefits that, people complain, are often lacking in today's marketplace: good service, customized suggestions, and low prices. As I will show in chapters 6 and 7, activities connected to this strategic management involve a balancing act that can have alarming consequences for consumers.

One part of the balancing act involves the perception of risk. Firms have a strong interest in reducing customers' fear of the web so that they will shop online and share information with trusted commercial sources. At the same time, consumers' fear of the marketplace serves to keep the already powerful firms on top. Calibrated and pitched correctly to the right customers, marketing causes people to place their trust in a known retailer, manufacturer, or media firm rather than in a new firm that wants their business. The message is "better to give Comcast and Bloomingdale's the information they want in order to be treated right than to share it with new firms that might offer better deals but are relatively unknown and so shouldn't be trusted."

But even the largest and most reputable firms develop strategic logics—assumptions about how to move forward in the uncertain digital environment—that undermine customers' trust as they promise to enhance it. They do that by increasingly using, or planning to use, customer information for profiling designed to lead to customized advertising, content, and pricing. These are activities about which, Annenberg research shows, the overwhelming number of customers who give firms information have little understanding. Executives hesitate to tell customers what they are doing, for fear of increasing customers' sense of harm and losing their confidence. As a result, large portions of the population are moving into a new age of media and marketing with high levels of nervousness about firms' knowledge about them, but with no understanding of the real costs of giving up personal information, or even of the kinds of data many companies have and can legally use. In this book, I will explore how we got to this point and where media and marketing seem to be going. I will examine the new industrial logic that is emerging as media and marketing executives try to meet fundamental challenges, the activities that are emerging as a result of the logic, the social implications that seem to flow from them, the resistance that individuals and organizations are attempting because of their social concerns, and how firms are responding to these reactions.

::

In chapter 2, I will begin to put the current feeling of crisis in the advertising industry in historical perspective. I will show that anxieties about consumers' power to evade advertising are leading marketers to urge Americans beyond "keeping up with the Joneses" and into a new era of thinking about social desire and what they must do to get what they want. I will investigate the conditions that led to this transformation, its relationship to the feeling of crisis that advertising and media practitioners feel about their work, and the strategic logic that has led them to turn to areas of the business that mainstream advertising personnel had looked down on in previous decades.

In chapter 3, I will trace the roots of product placement and direct-response advertising. Direct practioners, whose job it is to draw specific replies to an ad, were the acknowledged kings of marketing communication until the 1920s. The importance of product placement was much recognized in moviemaking and in broadcasting through the 1950s. Both businesses continued to serve clients vigorously after their most celebrated periods. Yet the sense of their centrality to marketing communication declined among mainstream media and ad executives. It was restored in the 1980s and 1990s, when mainstream approaches based on image advertising didn't seem to offer solutions to many of the problems of the digital environment. Marketers then turned to direct-response advertising and product placement, which were newly rehabilitated and which were trendy for other reasons. As these activities moved the new imperatives of marketers forward, it became clear that marketers and media practitioners were pressing for the most profound transformation of media and marketing's relation to American life in more than 100 years.

In chapter 4, I will show how that transformation took shape in the online environment. The internet, now the most interactive of electronic media, has become a test bed. Marketers have built on the traditions of product placement and direct response and have transformed both. Much of the time, their recipe has involved trying to take charge by attempting to inculcate a strong sense of brand trust while gathering information with which to decide whether a customer is worth engaging in customized digital relationships. I will trace the development of a database-marketing approach to consumers that includes six activities: screening for appropriateness, interactivity, targeted tracking, data mining, mass customization, and the cultivation of relationships. But in the course of ten years, attempts by marketers to exploit customer information in the digital world have led to an environment of mutual suspicion. Direct marketing to gain consumers' trust takes place while advertisers fear consumers' power and attempt to use data in more ways than they want customers to know, so as to profitably counteract their power. In that sense, trust and the undermining of trust often go hand in hand.

In chapter 5, I will show how some of the six database-marketing activities are beginning to emerge in a radical transformation of the home audio-visual environment. Influential executives are beginning to accept the basic logic of bringing a database-and-response mindset to television as much as they have accepted it online. They know that television technology is not yet advanced enough to combine screening for appropriateness, interactivity, targeted tracking, data mining, mass customization, and the cultivation of relationships in one advertising application. Cable, satellite, and even telephone companies that supply television signals are, however, testing many aspects of these activities. They are convinced that if they don't understand and apply new data-driven models, their competitors will. Up for debate is how far each of these approaches should and can go in interacting with specific members of the TV audience. A related question is whether these approaches should be dedicated to the separate advertising space surrounding television's entertainment narratives or whether interactivity and database targeting should also apply to product integration. One way to proceed is to encourage viewers to interact with sales messages that interest them. Another is to increase the customization ability of television's commercial messages and programming by using databases to vary the contents of programs and commercials in accordance with knowledge about households and even individual viewers.

In chapter 6, I will show how firms are learning how to apply the techniques of database marketing to people on the go, particularly in stores. Major developments in the use of database marketing at the retail level are paralleling the developments in digital media charted in earlier chapters. With changes in the commercial environment, with new information technologies, and with new analytical techniques, fundamentally new ways for stores to think about and treat customers are emerging. Like the new media regime, these are built on data mining, screening for appropriateness, targeted tracking, interactivity, mass customization, and the cultivation of relationships based on those activities.

The aforementioned developments support and are supported by major changes in digital media. Retailers and their suppliers are learning to use the internet, interactive television, mobile telephones, and other consumer-driven interactive technologies to find new customers, gather information on new and old ones, and reach out to consumers with advertisements and content rewards that are increasingly tailored to what the databases know. They aim to create customized environmental surrounds that inspire trust and return business. As part of this process, retailers are increasingly placing pressure on desirable customers to identify themselves if they want to be treated especially well. Moreover, a noted consultant stated, the word is getting out that the best customers, placed in the best niches, will get the best deals. Speaking about the incentive people have for identifying themselves to stores, he stated: "People not in the right segments will be left behind. They will not have as rewarding an experience."

Marketers will often exploit information about customers in more and different ways than they expect. Having shown in chapters 2–6 how this activity takes place across a wide range of venues, in chapter 7 I will explore the extent and the nature of public resistance to these activities. Concerns are flowing from various quarters, and executives are parrying them with public claims and behind-the-scenes lobbying. The claims have serious holes, but marketing and media practitioners are fortunate to have the social environment as an ally in keeping the flaws mostly hidden and the public stress levels controllable. Consumers have little knowledge about retailers' power over information, government agencies focus mostly on scams and narrow meanings of privacy, and advocacy groups' views of database marketing do not get much coverage in the popular press.

A major goal of marketers and media outlets is to persuade customers who fit a desired profile to give up information in exchange for being considered special and not having their information abused. One can imagine marketers knitting the marketing and media developments discussed in the previous chapters into an even more integrated version of database marketing. The most elaborate possibilities are not yet happening; however, the industrial logic points in that direction, and the technology is evolving to make it eminently possible.

These developments may well result in a marketing-and-media world that varies with what niche marketers put an individual in.

In chapter 8, I will step back to consider the social meaning of database marketing. Extending the idea that database marketing is beginning to engender new forms of envy, suspicion, and institutional distrust, I will argue that it works against a sense of social belonging and engagement. Finally, I will suggest ways to counter the trends in ways that encourage public knowledge and force media and marketers to be more open about how they get and use that increasingly valuable currency, personal information.

2 :: Confronting New Worries

"I know that half of my advertising budget is wasted, but I'm not sure which half." Though it is one of the most commonly quoted epigrams in the advertising industry, writers disagree on who said it first. In the United States, they typically give the credit to the late-nineteenth-century Philadelphia retailer John Wanamaker. U.K. marketing trade magazines often name the late-nineteenth-century soap manufacturer Lord Leverhulme. One advertising chronicler attributes the thought to that era's publisher of the *New York Times*, Adolf Ochs.[1]

It is significant that the sentiment has been associated with a manufacturer, a retailer, and a media owner but not with an advertising agency executive. Agency people often grumble that manufacturers, retailers, and media owners have a strong interest in blaming all sales difficulties on them. The epigram's implied criticism of their work raises many agency folks' hackles. Winston Fletcher, chairman of a well-known British advertising agency, let loose with this critique:

What on earth could it mean? That half of all the commercials transmitted have no effect, while the other half do the business? That half the people who see each ad ignore it while the other half reach for their pocketbook? That the first 15 seconds of every commercial are dud, but the second 15 seconds are real goers? Or maybe it is every alternate second, or every alternate nanosecond?

And why pick on advertising? Why has nobody ever wittily aphorised: I know half the legal fees I pay are wasted but I've no means of knowing which half'? And what about half the market research you carry out? And half the documents you read? And half the product development that's carried out? Or half the sales calls that are made?

Now we're getting somewhere. Salesmen rarely, if ever, know for sure, beforehand, whether or not the punter will buy. That doesn't mean the unproductive calls are wasted. Without the unproductive calls there would be no productive calls. Every salesman knows a 100 percent strike rate is a mythical beast: his job is to maximise that percentage.

It's the same with ads. . . . But nothing will stop daffy people from saying it. Well if you're one of them, please keep in mind in future that you're talking utter crap. Thank you.[2]

But while comments such as these intend to dismiss the aphorism, they actually betray a deep anxiety that is at its heart. There is no denying that, ever since the start of their industry in the nineteenth century, advertising organizations have been deeply worried about finding efficient ways to persuade people to buy their products. Much of the history of the modern advertising industry has involved the advertisers, media, and agencies struggling with each other and with consumers to accomplish that goal. And while advertising practitioners have understood that a perfect solution is unachievable, that hasn't stopped them from being frustrated about it and searching for the best new answers.

In this chapter I explore the approaches that the advertising industry developed in the nineteenth and twentieth centuries to profitably persuade audiences. Their work profoundly influenced advertising messages, the media and society at large. It built into the media system a new ideology about consumption and about the responsibilities of advertisers and the public to each other. Still, as that oft-repeated phrase from Wanamaker (or Lord Leverhulme or Adolf Ochs) implies, gaining control of a crucial part of the twentieth century's cultural apparatus didn't eliminate the worries with respect to reaching and persuading people efficiently. What it did do was create an equilibrium in the relationship among media, marketers, and the public that allowed advertisers to feel that they were operating in a world that was to some extent under their control. They could tell themselves they could be sure about the efficacy of most if not all of their advertising monies.

That equilibrium seems to be falling apart as new media raise the ghost of large-scale audience unpredictability and unresponsiveness. Marketers, bereft of their comfort zone, are scrambling to develop alternatives. Those alternatives promise to reshape the media and society as surely as the activities at the turn of the twentieth century did. To understand where marketers and media practitioners are trying to go and how, it is first useful to track where they are coming from.

::

The advertising industry was one of several forces emerging in the decades after 1865 that changed Americans' approach to the consumption of goods. With the end of the Civil War came explosive changes in industry, commerce, and population. The manufacturing capacity of the United States increased sevenfold between 1865 and 1900. Railroads expanded, and migration to the West accelerated. New mines were discovered and put into operation. Industry began to move from the East into the Midwest. New factories drew workers from American farming communities and foreign countries to New York, Chicago, Philadelphia, and other cities. Between 1870 and 1900, the United States doubled its population and tripled its number of urban residents. The country was changing rapidly from an agricultural to an industrial economy.[3]

These changes had consequences for the kinds of goods people could buy and for how they could buy them. Urban factory workers had no time or ability to grow their own food or make their own clothes from scratch. Many of the factories created products that had been made by hand, often by the user's family, only a few years before. Other plants turned out things—toothpaste, corn flakes, safety razors, cameras—that nobody had made previously.

Many Americans first found out that they could buy these things—in fact, that the goods even existed—through two channels: the department store and the advertisement. Historians date the first American department store to 1826 (in Duxbury, Massachusetts); they point out that advertising in the United States started with the first colonial merchants.[4] Still, department stores and advertising developed through the late nineteenth century in ways that crucially reflected and affected major changes in American society's approach to goods.

The historian Daniel Boorstin called department stores "Palaces of Consumption."[5] Stewart's in New York, John Wanamaker in Philadelphia, Jordan Marsh in Boston, and Field, Leiter & Company in Chicago offered a wide range of merchandise, including clothing, small household wares, and home furnishings, for public view and purchase. It is difficult today to understand the novelty that these places brought. Traditionally, stores placed in public view only those items that the merchants believed everyone could afford. They invited relatively wealthy individuals to visit special areas away from the rabble where they could experience more expensive goods. Then, too, there were merchants who dealt only with the rich. That anyone could gain free admission to see a wide range of expensive and cheap goods for sale was a novel idea built into the department store model. Clearly, not all those who saw the goods could afford them. Still, they could see reflected in those goods worlds previously closed to them, and they could dream of future possibilities for themselves or their children. The stores also democratized price. The policy of fixed fees, begun by Stewart's in the United States, was born of business necessity. Store owners could not trust their hundreds or even thousands of employees to haggle profitably with customers. Nevertheless, this nod to transparency fit well with the other elements of openness that department stores typically announced: free delivery, freedom to return or exchange goods ("satisfaction guaranteed or your money back"), and charge accounts. The message was that the stores wanted to maximize the number of people able to buy what they sold.

Many of the department store owners were convinced that they had to advertise to attract the hordes that would make their emporia profitable. Billboards, streetcars, and especially newspapers were the vehicles for bringing awareness of the new retail cornucopia to the masses. John Wanamaker was the largest local advertiser in the United States, and the most enthusiastic. Like other department store moguls, he hired his own advertising writers to write copy for newspaper ads that aimed to bring crowds of buyers into his store every day. By the late 1880s, he was up to two or three newspaper pages daily—costing between $300,000 and $400,000 a year—in what he called "common-sense" advertising.[6]

The "common-sense" label Wanamaker attached to his advertising was shrewd. This was a time when much of the advertising was associated with the far-fetched, typically fraudulent, and often dangerous claims of patent-medicine hawkers. Newspapers were filled with the stuff, often targeted to Wanamaker's chief customers: women. Wanamaker clearly wanted to give shoppers the sense that his ads were an extension of the straightforward presentation of the goods and their prices in his store.

John E. Powers, who wrote newspaper ads for Wanamaker in the 1880s, helped establish an honest, transparent print personality for the store through an approach that writers for department stores around the country imitated. Powers gave the impression of telling the truth about the goods to the point that at times he would denigrate them, on the premise that people would know what they were getting when they shopped at Wanamaker's. In one advertisement he noted that a stock of neckties would be reduced that day from $1. "They're not as good as they look," he wrote, "but they're good enough—25¢." To sell light waterproof jackets that the "rubber department" told him they had not been able to move, he announced in the next day's newspaper: "We have a lot of rotten gossamers and things we want to get rid of." The raincoats were sold out that morning.[7]

Roughly paralleling the development of fixed-price department stores and their advertising rhetoric was the spread of chain stores (A&P, Woolworth) and catalogue firms (Sears, Montgomery Ward). The chains tended to sell different goods and be less expensive than the department stores; the catalogue companies reached out to rural populations that couldn't get to the department stores or chains. The chains, the catalogue firms, and the department stores were all palaces of consumption. With their fixed-price polices, wide range of products, extension of credit, and emphasis on customer trust through satisfaction, they beckoned all sorts of people to their midst. It was an invitation to ogle, assess, and compare worlds of goods that had not existed for them just a few decades earlier.

The invitations affected not only consumers and retailers but also manufacturers. They found themselves competing fiercely for store space and popular interest with other companies that made similar items. They therefore had to come up with ways that would persuade stores to carry them and consumers to buy them. The savviest companies found the answer to both by advertising to broad audiences. One goal of the advertising barrages was to persuade consumers to ask retailers for the firms' specific products. A no-less-important aim was to impress on retailers that those requests were coming and that merchants had to be prepared by carrying the goods in the first place.

Many of the national ads showed up in the popular magazines that were finding their ways into millions of American homes by the beginning of the twentieth century. The magazine entrepreneurs Frank Munsey, Samuel McClure, and John Brisbane Walker had replaced the poetry, literary criticism, and formal essays of established periodicals with light fiction and journalistic reporting of contemporary events. More significantly, they had priced their issues (and their subscriptions) far below the costs of traditional magazines. The aim was to attract members of the growing middle and working classes who had not previously bought periodicals. The publishers' innovation was to place most of the charges for production and distribution costs not on the readers but on advertisers who were interested in reaching them.

In 1894, when *McClure's* was about a year old, its July issue carried 31 pages of ads. By 1901, a typical issue had 105 pages of ads. Most of the national advertisers of the day were represented. The June edition had ads for Ivory Soap, Welch's Grape Juice, Hires' Root Beer, Wheatena, Kodak, the National Biscuit Company, Elgin Watches, and Johnson's Wax, among others.[8] At least as important as the presence of these goods was how they were presented. Trademarked characters, logos, and bold copy were intended to inculcate memory and to stir action regarding the registered names, and the synergy of advertisements and in-store packaging recalled ads that aimed to give goods personalities so that consumers would choose those specific products, not those of competitors, even if the price was higher. Manufacturers hoped that stores would anticipate that customers would ask for those advertised brands and would feel obligated to carry them.

Kellogg's played out the idea quite directly in *Munsey's* in the early 1900s with an advertisement featuring a fashionably coiffed young woman sternly stating to an unseen grocer that she did not want any substitutes: "Excuse me—I know what I want, and I want what I asked for—TOASTED CORN FLAKES—Good day." Another Kellogg's ad of the era showed a clown, a dog, a cat, and a baby howling "The Song of Imitators—We're just as good as Kellogg's." The ad went on to caution that "there are none so good and absolutely none are genuine without this signature, W. K. Kellogg."[9]

Fine-tuned language that reflected the competitive flavor of the new commercial environment was increasingly a must for companies reaching out to the buying public. Until the last decade of the nineteenth century, merchants or manufacturers typically wrote the ads, even if they had hired an advertising agency. The first ad agencies were founded in the 1840s. Their purpose was to help store owners buy space in large numbers of newspapers or magazines. That changed as commercial competition heated up and ad agencies saw that having copy experts and art experts on staff would help them lure clients. By about 1910 the crafting of national ads had become the province of agencies.

The activity attracted and created stars. Well-known writers such as Bret Harte and Artemus Ward penned material for Sopolio soap. Popular poets such as Oliver Herford and Madison Cawein composed ditties to promote Force cereal. At the same time, advertising practitioners known to have a gift for persuasion (often with experience in the patent-medicine field) could demand handsome sums. In 1899, the Lord and Thomas agency in Chicago paid the patent-medicine promoter John E. Kennedy $28,000 a year to write its client's ads. When Kennedy resigned in 1907, the firm hired another well-known patent-medicine copywriter, Claude Hopkins, for the even more princely salary of $52,000 a year.[10] Kennedy and Hopkins applied the techniques they had learned hyping elixirs to clients as varied as Van Camp's beans, Goodyear tires, and Oldsmobile cars.

As advertising writing became an identifiable craft, clear styles developed—jingles, bombastic prose, and seemingly fact-based persuasion. Together, they presented consumers with a rhetoric of the importance of understanding the fashion, the utility, the quality, and often the price of products and services. The appeals complemented the messages consumers were getting when they visited retail establishments. It was hard to escape a collective message in the riot of commercial claims in the mass-oriented newspapers, magazines, and billboards: Consumption was the order of the day, and stores, catalogues, and ads were giving Americans an opportunity to survey the landscape, evaluate the goods, and look forward to purchasing what they wanted and could afford.

::

All this was happening with blinding rapidity. In just a few decades, American marketers had changed their environment fundamentally to allow them to reach out to large numbers of potential customers. Yet they still faced a major barrier to selling. For many Americans at the turn of the twentieth century, the new landscape raised strong ethical concerns about material goods and envy that made it difficult for them to accept marketers' blandishments without great hesitation. Assuaging the religious anxiety that Americans (particularly women) felt regarding this issue became a central concern for advertisers and the media that supported them. They helped to cultivate a new perspective encapsulated in the phrase "keeping up with the Joneses" to assure people that it was not only acceptable to be envious, it was socially useful. As marketers faced criticism for helping Americans channel their desires, they tried to persuade them that envy, in an age of plenty, wasn't bad.

Of course, envy of the wealthy by the less wealthy has probably always existed, despite religious edicts against covetousness. Proscriptions notwithstanding, one common way to channel envy was to imitate the wealthy with less expensive materials. That often took enormous effort. Women who wanted symbols of an elevated social status but could not afford to buy them had no choice but to search for the right materials and reproduce the coveted objects from memory.[11]

Imitating the wealthy got a bit easier after 1863, when the Butterick company and magazines such as *The Delineator* and *Harper's Bazaar* began to offer patterns that allowed more precise modeling of upscale styles. Emulation was made even easier by department stores, chain stores, and catalogues, which enabled women to purchase versions of upper-class clothing styles and home furnishing at middle-class prices. Envy was especially important in the case of home furnishings. The Victorian notion of the middle-class home was of a place apart from the cacophony of the outside world, a place where a husband could re-energize before going out into the workplace fray and where children could grow up in safety, nurtured by the kind of love that was so often absent from the urban setting. Yet the home was quite a bit more porous to the outside—to neighbors, relatives, salesmen, delivery people—than this belief suggested. The sociologist Thorstein Veblen observed at the time that middle-class householders understood that the home was a means of displaying the male breadwinner's earning power.

By visiting stores, looking at catalogues, and reading magazines such as the *Ladies Home Journal* (which published articles with such titles as "The Ideal Kitchen" and "Looking into Other Women's Homes"), middle-class women could easily note the aspects of the most fashionable clothes and homes that they were lacking and what that said about their social status. The Industrial Revolution's focus on commercial goods created an arena for social envy. At the same time, the stores and catalogues that helped fuel the desire also presented profit-inspired antidotes. Middle-class individuals could now buy mass-produced versions of the clothes, rugs, and other accoutrements that the wealthy had.

The wide availability and display of domestic goods *increased* envy, however. Mass production had made it possible for women with widely differing incomes to purchase similar rugs, pianos, dresses, and other items for themselves, their families, and their homes. But, as the historian Susan Matt notes, "status-conscious women who worked to resemble social elites expressed their discomfort and resentment when women lower down in the class hierarchy imitated them." Manufacturers and retailers had to consider, then, that "members of the middle class not only imitated the styles favoured by those above them on the social scale, but they also worked to separate themselves from the social classes below them."[12] The result was an ever-increasing stream of products, styles, price points, and claims that encouraged customers to believe that their purchases would make them stylish and distinctive.

While these developments confronted Americans with a cornucopia of goods, they also generated nervousness among nineteenth-century elites about the acceleration of envy-driven materialism. Essayists, clergy, and editorial writers railed against covetousness as a grave sin that could lead women and their families to ruin. Drawing on religious views of a strictly hierarchical world, they urged people to accept their social positions as the natural order of things. Editorialists and ministers repeated that women who wore imitations of high fashion were insincere dissemblers. "You have been placed in a certain position of life," wrote Edward Bok, the editor of the *Ladies Home Journal*, in 1891. "Instead of trying to cover your real position with sham, why not adorn it and make yourself envied for your own qualities if not for your possessions?"[13] Bok and many other thought leaders advocated contentment as the antidote to envy. In newspaper columns and in magazine stories, they warned women that excessive and competitive spending threatened not only a family's finances but also its stability. A 1914 book titled *The Girl That Goes Wrong* claimed to offer evidence in tales about women who suffered egregiously—even became prostitutes—because they had yielded to envy.[14]

By 1914, however, this view of material envy as pathological was fading from the popular culture. Instead, a number of considerations were coalescing to encourage strong compulsions toward upward mobility and the fashions that went with it. One force driving this development was the widespread perception that the United States was becoming a society of abundance with enough resources to satisfy all who worked to get what they wanted. Another driving force was the replacement of the medieval Christian view of fixed social positions with a dynamic "survival of the fittest" perspective associated with Charles Darwin's recently published theory of evolution. More and more people believed that it was their right, perhaps even their obligation, to strive for a better position in society.

Corporate executives as well as economists, social reformers, and journalists trumpeted this new philosophy as the nineteenth century's conservative opinion leaders died or faded from the scene. The rising influentials argued that competitive business instincts and widespread spending were helping the United States move forward. Envy and discontent were therefore positive social instincts.

The new advertising-supported magazines certainly had a vested interest in pushing the idea that commercial striving was the American Way. Stories and columns carrying this message meshed with the ads in the periodicals as they implicitly encouraged their readers to try out the products. Beginning around 1910, articles in popular magazines explicitly rejected the idea that women who dressed "above their station" were insincerely dissemblers. A 1911 column in the *Ladies Home Journal* contended that "every woman ought to be dressed just as beautifully as she can possibly afford to be, without risking bankrupting her husband—and she need not worry about this latter consideration." The argument was that by acquiring the correct status symbols in clothes and home furnishings a wife could help her husband look good to his bosses and so advance in life. Few if any men, the writer added, "have been bankrupted by their wife's extravagance." That calumny, he said, "is chiefly a belated echo of the old whine in the Garden of Eden."[15] Here was a popular magazine invoking the Bible to support envy-driven consumption. The rhetoric reflected a more general reality: By the 1910s, the mainstream media were reassuring their audiences that buying things out of envy or desire for social escalation was simply "keeping up with the Joneses." The phrase came from a popular comic strip of the decade, and it enjoyed great currency by the 1920s. As used during that period, it was a matter-of-fact expression of the view that envy and imitation were ordinary social instincts.[16]

To advertisers, the notion of "keeping up with the Joneses" was a godsend. As competition for customers grew, a philosophy that placed the desire for goods near the center of family life meant that people would see a broad array of commodities as acceptable for purchase. The growing centrality of material desire and upward mobility also helped to justify the high-profile presence of consumption-oriented advertising in mass-audience magazines and newspapers.

That didn't mean it was all clear sailing for advertisers. Muckraking magazine and newspaper stories about the advertising of patent medicines and about other frauds raised questions about all ads and the ethics of the media that carried them. Worried about an emerging pattern of government intrusion and about their loss of consumer credibility, major advertisers and media outlets assured their readers that the advertising business would help them sort out who was honest and who was not. Industry groups drew up "Truth in Advertising" codes and formed an organization to help wronged consumers. (Later it would be known as the Better Business Bureau.) To enhance their own credibility, several important magazines and newspapers announced that would take responsibility for claims of ads they carried. The money-back guarantee was another tool in the struggle to gain public confidence and to sustain the momentum of material desire. The magazine *Collier's* described the promise as "the most powerful of moral influences being exerted today."[17] Edward Bok, an enthusiastic convert to the new style of marketing, took up a more general defense of the messages of desire. "The fact must never be forgotten," he wrote, "that no magazine published in the United States could give what it is giving to the reader each month if it were not for the revenue which the advertiser brings the magazine."[18] Advertisers succeeded quite quickly in getting Americans to accept ads in exchange for low-cost, high-interest media content. In the 1920s, politicians, regulators, and even some trade magazines spoke out against radio advertising as too intrusive in the home. The advertising trade magazine *Printer's Ink* editorialized that "the family circle is not a public place, and advertising has no business intruding there unless it is invited."[19] Yet radio stations and their allies decried schemes to support the radio system through taxes and simultaneously arranged to sponsor broadcasts featuring popular entertainers. As one chronicler generally sympathetic to the advertising industry noted,

People accepted with complete equanimity the pushing back of the walls of their homes [by radio] to world horizons and, without protest enough to bar their way, the clamoring push of salesmen for products of all kinds into their living rooms and boudoirs. Instead of reading advertising if and when they pleased in magazines and newspapers, on car card or billboards, they knew its din from "Cheerio" in the morning to the last adult bedtime story at night. They tolerated, if they did not enjoy, the enforced intrusion as inevitable, seeing it as the price they had to pay for heady pleasures.[20]

Despite their success in insinuating themselves into the most private areas of Americans' lives, advertisers and ad agencies remained nervous about the twin job of getting people's attention and impelling them to buy. Stories in and out of the trade noted that people talked, visited the kitchen, or used the bathroom during radio's commercial breaks. Advertisers tended to view this sort of distraction as nothing new. They knew that for centuries newspapers and magazines could not guarantee that their readers would look at, let alone read, every print advertisement. Similarly, advertisers understood that radio stations and networks had no way to force people to keep listening when commercial messages came on. Advertising agencies, in fact, saw garnering people's attention through compelling ads as one of their primary challenges. They were led in this direction by Daniel Starch, who in the early decades of the twentieth century pioneered ways of measuring the readership of a print advertisement. Advertising practitioners carried the notion over to radio and television. In the absence of proving a direct relationship between commercial and purchase, "recall" became a surrogate for a commercial's success. Copywriters and art directors made high recall an important value in a commercial's creation.

In their attempts to get good recall and to encourage purchasing, ad executives turned to the emerging science of desire. In the early 1900s, academic psychologists were beginning to lay out propositions about how human instincts could be exploited to elicit material wishes on the part of consumers. In 1911, Walter Dill Scott of Northwestern University noted that "goods offered as means of gaining social prestige make their appeals to one of the most profound of the human instincts."[21] John Watson, a psychologist at Johns Hopkins, went further. He offered the possibility that even if individuals did not see goods as attractively related to social prestige or other desirable goals, they could be made to feel that way through conditioning. A president of the American Psychological Association and a founder of behaviorist psychology, Watson went to work for the J. Walter Thompson ad agency in 1916, while he was still a professor at Johns Hopkins. His goal at J. Walter Thompson was to develop techniques that would allow him to condition and control the emotions of fear, rage, and love to improve the effects of advertising on the consumers. In 1920, after

Johns Hopkins fired him for having an extramarital affair with a graduate assistant, Watson turned to developing the psychology of advertising full-time, first at J. Walter Thompson and later at other firms. His insistence that advertising could channel emotions in support of product purchases made an important mark on ad-agency researchers and "creatives" from the 1920s through the mid 1940s.

Influenced by Watson's work as well as by Freudian psychology, advertising practitioners began to put a scientific aura around what some copywriters were already doing without any fancy theories: displacing their own angst about reaching and persuading potential customers by trying to create worries among the public that would benefit marketers. The aim was to increase consumers' angst by stimulating them to fret about their inadequacies or to envy the social success of others. This was a turning point in the way advertisers thought about their audiences. The ethically questionable techniques used to advertise patent medicines had preyed on problems people knew about. Now advertising people became intent on using science both to find out the best tensions to associate with particular products and to present specific products as solutions to problems people didn't know they had.

The Lambert Pharmaceutical Company sold a disinfectant mouthwash called Listerine on the basis of the notion that people's lives were being destroyed by bad breadth but they didn't know it. "Often a bridesmaid but never a bride," announced one 1920s headline. "Even your best friend won't tell you," another chillingly pronounced. During the economic depression of the 1930s, the Gillette razor company bought space for a dramatic photo of a disconsolate man in his late twenties or early thirties talking to an equally sad woman, clearly his wife. On the man's face, dark stubble could be seen. "I didn't get the job," he said. In the 1940s, as I have already mentioned, children listening to the national radio program *Little Orphan Annie* were told in a long commercial that they surely would want to be among the first among their friends to have the 1941 edition of the Little Orphan Annie drinking cup, which they could get by sending two Ovaltine labels to a certain address.

By the mid 1920s, the developments had become an obvious trend. The sociologists Robert and Helen Lynd saw it during their now-classic exploration of life in a small Indiana city during the 1920s. They observed that, unlike ads of a generation earlier, advertising of their era was "concentrating increasingly upon a type of copy aiming to make the reader emotionally uneasy, to bludgeon him with the fact that decent people don't live the way he does."[22] Here was "keeping up with the Joneses" put in a darker, more angst-inducing way. Advertising's critics lambasted this side of the industry's activities as part of a general pattern of problems that commercialism had brought. In 1934 Robert Lynd wrote: "The consumer stands there alone—a man barehanded, against the accumulated momentum of 43,000,000 horse power and their army of salesmen, advertising men, and other jockeys. He knows he buys wastefully . . . that his desires and insecurities are exploited continually, that even his Government withholds from him vitally important information by which both it and industry save millions of dollars annually."[23] Not surprisingly, marketing and media executives disagreed. They argued that attacks on advertising during the 1930s had led to strong policing of claims and more honest pricing than ever. Riding herd over the industry, they said, were research-and-advocacy organizations such as Consumers Union as well as several federal agencies with the right to investigate and restrict advertising's excesses. Offenders caught by these agencies received lots of attention in the press. Yet, supporters of the advertising business argued, the great majority of ads were found to be acceptable in the eyes of a major government watchdog, the Federal Trade Commission. Between June 1941 and June 1942, the FTC examined 362,827 print ads and found that only 20 percent of them carried false and misleading representations. Of the 1,000,450 radio commercials the FTC examined, only 2 percent were found to be false and misleading.[24]

Responding to complaints by historians and social philosophers that advertising practitioners were encouraging waste, the advertising industry's supporters brought out academics to argue its social value. Of these academics, the one most often quoted was the Harvard University business professor Neil Borden, who in his 1942 book *The Economic Effects of Advertising* wrote: "Advertising's outstanding contribution to consumer welfare comes from its part in promoting a dynamic, expanding economy. . . . In a dynamic economy . . . advertising . . . is an integral part of a business system in which entrepreneurs are constantly striving to find new products and new product differentiations which consumers will want."[25] Critics sneered that much advertising didn't fit such high-minded goals, and the back-and-forth argumentation showed that the combatants disagreed on more than economic theories and facts to back them up. Up for grabs in the debate was a vision of the role that commercial-driven materialism should play in American society in the coming decades and how it should be regulated. Increasingly, many powerful federal policy experts of the late Depression era sided with Borden about advertising's role in encouraging useful competition. They also damped down activist government regulation of the industry. By the 1940s, business leaders had helped to shape most government expectations of them so that legislative responses to consumer activism would no longer be the primary order of the day. Instead, policy makers saw consumers' materialism—the collective buying power of the public—as the force that would move businesses to act in the public interest. This idea was that consumers held the present and future health of the American capitalist economy in their hands, and that companies would learn that they had to tread carefully and honestly if they wanted long-term success.

Coming out of World War II, then, the federal government's position was that the Americans' materialist drive was not a problem caused by marketers but a solution to potential marketing abuses. Advertising and media executives picked up on the idea that consumers were voting collectively through their purchases and attendance. The notion affected the way they talked about the Nielsen ratings of television shows. Based on diaries and meters that audited the viewing habits of a sample of American homes, executives used the ratings as the primary considerations in pricing commercial time. When intellectuals derided commercial television as the height of homogenized "mass culture," programmers of the 1950s and the 1960s defended their ratings-driven selections as "consumer sovereignty." Ratings, they argued, were a kind of nationally representative vote for the popular culture that advertisers were delivering to the American people.[26]

It all signaled success for an advertising industry that had worked so hard to create a structure that would address its angst about getting people's attention and persuading them to buy. The notion of "keeping up with the Joneses," the creation of messages that tried to displace marketers' angst with consumers' angst-driven envy, advertising people's supporting the media system to circulate those messages—all that had become accepted as parts of American culture. In a bit less than a century, the nation had become what Lizabeth Cohen calls a "consumer's republic," a society bound practically and politically to the idea of consumption.[27] Advertising and stores were the society's sometimes-annoying cheerleaders and guides. Americans now had broad access to goods and knowledge about them. Ads, public malls, and known pricing allowed them to "comparison shop" for an astounding array of products and services. Instalment plans and credit often allowed many to buy more than they could afford.

Marketers spent a lot to tell people about it, and they still worried that they didn't know quite what they got out of it. An extreme example of audience inattention in television's early days relates to the wild popularity of Milton Berle's television show, *Texaco Star Theater*. It seems that in Detroit so many people flushed the toilet simultaneously during the sponsors' segment that the water pressure dipped noticeably.[28] Yet the conventional wisdom was that, despite some inattention, most people did see advertising messages, and advertisers got more out of the advertising than they spent on it. Marketers believed that knowledge and control were generally on their side. Though customers could compare prices, ultimately manufacturers and retailers knew their margins and ingredients and often kept them close to the vest. Moreover, by funding the media they were purchasing the attention of the largest audience in history. Advertisers' most basic message to the nation had been honed through the first half of the century: "We'll pay for the content. You just pay attention to the ads—and buy."

::

For about four decades after World War II, that formulation and the industry routines that addressed it kept advertising practitioners' anxiousness about reaching audiences and persuading them audiences at levels that were acceptable to agencies and their clients. After 1945, companies that wanted to advertise their goods and services nationally found a well-established system waiting to support them. They could hire agencies experienced in audience analysis, media buying, and ad creation. They could buy time or space on mass media that existed to please sponsors and aimed at large audiences. They could plug into ratings operations that claimed to audit whether the audiences they were paying to reach actually showed up. And they would discern government regulatory behavior that, despite fits of judicial, legislative, and executive anger, was predictable enough not to interfere with the basic business of buying and selling mass audiences.

The broadcast television industry was the rising advertising star that set the tone for the next 35 years. Launched in the late 1940s on radio's model of sponsor-supported "free programming," television reached 86 percent of American households by 1960.[29] Television's enormous audience-drawing power forced other media—especially radio and magazines—to turn to targeting slices of the American population in order to be useful to advertisers. By 1985, though, a stream of new technologies led advertisers to worry about television's power to deliver America to their commercials. Television and marketing executives initially thought that targeting would be the solution to the changes that were convulsing the businesses, but by 2000 advertisers and executives knew they were wrong. The advertising industry's historical angst began to erupt beyond bounds that agencies and clients found tolerable. Marketing and media leaders worried that fundamental changes were needed. Their initial belief that targeting could fully address the grave challenges the new media held for advertisers came from a failure to recognize two features of the technologies emerging through the 1980s and the 1990s: (1) Many new devices allowed for audience division. (2) Some also allowed the audience to escape advertising messages. Executives quickly recognized the technologies of division, but they were slower to understand the challenges posed by the technologies of escape.

The audience-fragmenting implications of new technologies emerged with the television age. According to the Nielsen ratings company, on a typical evening in the 1970s, the CBS, NBC, and ABC television networks together reached 90 percent of American households with their sets on. On high-viewing nights such as Sunday, that could easily translated into near 60 percent of all homes in the country. This unprecedented audience-gathering ability shook the magazine and radio industries to their core. Looking at plummeting network radio ratings, national advertisers inferred that network radio was losing much of its broad-based audiences to television during the 1950s. As for the mass-circulation magazines, many of them continued to gather huge numbers of readers by offering them extremely low subscription costs. While that was great for advertisers before television, advertising people now noted that via the home tube they could reach the same kinds of large, diverse audiences they purchased through magazines such as the *Saturday Evening Post, Collier's*, and *Life* at comparable costs. They judged television better, though, because it had the audio-visual impact of motion pictures.

As the mass-oriented periodicals and network radio went down in flames, new target-oriented industries arose from the ashes. Magazines and radio stations that called out to specific audience categories—by gender, race, age, lifestyle tastes—became the norm. Many advertisers found these sorts of division useful. Increased competition was leading manufacturers to design ways to differentiate products so that smaller and smaller numbers of a product could be made and marketed profitably to certain segments of society. Spurred by the need to learn about the niches that might use the products, market research firms were coming up with new ways to differentiate parts of the population of interest to manufacturers, retailers and media. Items that seemed basic suddenly were changed to fit various lifestyles. "In the old days," a Procter & Gamble executive noted in 1994, "Tide was one big brand. It stood for clean, white clothes and all women 18 to 49, whether they had kids, or didn't have kids, washed their clothes [with it]. But now, you have Tide with Bleach, Tide Ultra, Tide Unscented. And each of these brands are still targeted at women 18 to 49, but they are targeted at differences between segments of women 18 to 49."[30]

It was in this manufacturing and marketing environment that widespread talk grew among marketers in the late 1970s about using television to reach different audiences in the same way they were doing with radio and magazines. Cable was the first technology to spur advertisers toward thinking about changes that they would have to make in their approach to the home tube and its viewers. Cable had been around since the 1940s as an antenna service for rural homes that could not otherwise get television, but until the 1980s its use in American homes had been negligible. In the mid 1970s, however, U.S. government policy changed to encourage the urban spread of cable and the distribution of television programs by satellite. Both developments led directly to new nationwide cable networks and discussions among advertisers of a hundred-channel universe within a few years.

The notion of so many channels initially scared marketers, but they interpreted the new media scene as a reflection of the fractionalizing, frenetic, and self-centered American society.[31] Targeting became the goal, and marketers and media personnel were confident that they could control the emerging environment along much the same lines that they had controlled the advertising world. "Keeping up with the Joneses" was no longer a popular phrase. Nevertheless, the idea behind it remained at the core of marketers' strategies. Advertisers encouraged people to buy on the basis of being attractive to people like them. And they took for granted that if they used new market research techniques they could target the right people with the right commercials effectively.

They were slower to realize that technologies of escape were creeping onto the scene. It isn't as if the idea was entirely new in the second half of the twentieth century. Advertising people during the first half worried about angry publics that wanted to eliminate billboards as well as newspaper, magazine, and radio audiences who did not recall ads. In 1906, Lee DeForest invented the audion, a device that made it possible to amplify radio signals. DeForest vehemently opposed advertising.[32] In a 1930 article in *Radio News,* he described how a wireless remote control might be developed with which "the long-suffering radio user" could "instantly assassinate the advertising announcer and allow the set to resume its musical outpourings when the story of the tooth-paste or furniture salesman is terminated."[33] No such device was sold commercially, and the notion seems to have caused no advertisers to lose sleep.

Yet the idea of using technology to escape advertising revived slowly in the 1970s. Viewers' ability to reject the ads while watching the shows received a boost with the introduction of the hand-held remote-control apparatus. Because the Nielsen television ratings company's technology could not at the time note quick channel switching during commercials, the advertising trade did not pick up what the technology was starting when it was introduced. As Nielsen meters became sophisticated enough to track channel flipping, advertisers noted with consternation that people with remotes were using commercial breaks to see what other programming was available. The spread and proliferation of cable television channels in the 1980s and the 1990s gave viewers even more incentive to "surf" during commercials. Newspapers and magazines even began to spread the word that young people were using the remote to view a number of programs on different channels at the same time. Clearly, that put advertising material in danger of being zapped in favor of the multiple shows themselves.

Media planners who looked upon the remote control with some nervousness received an added jolt with the rise of the videocassette recorder beginning the late 1970s. Optimists saw the VCR as a vehicle for time shifting, so that people who could not view a particular program at one time might view it at another. Pessimists pointed out the VCR remote allowed users to run rapidly through commercials, thus invalidating advertisers' expensive purchases of time. A great discussion ensued throughout the advertising industry about what to do about this. One response was to lengthen the time a company's logo was shown in an attempt to ensure that viewers racing through the commercial would at least register that.[34]

Other reactions by advertising practitioners to the remote control during the 1980s and the 1990s reflected a mixed sense of blaming themselves and the television industry. Advertising executives who looked to their own industry argued that audiences simply did not find commercials interesting enough. They exhorted creative types that making commercials that people would want to watch would go a long way toward facing up to this technological challenge. Others weren't so sure. They blamed the television industry for creating so cluttered a television environment that viewers would try to escape it now that they had the technology to do so. Network executives replied that the presence of rabid competition from so many audio-visual channels was forcing them to increase rather than reduce publicity for their shows. In response to advertisers' anger, however, the major broadcast networks did make faint efforts to cut back on their on-air own promotional activities. Instead of stand-alone ads, for example, they thrust publicity for upcoming shows onto the credits at the end of programs.[35]

For advertisers, though, the sense of increasing commercial cacophony that the audience hated remained, and was spreading to new media. With even greater alarm, they noted that technologies aimed specifically at helping people escape commercials were appearing. Of particular concern were technologies that could stop unwanted internet advertising (especially pop-ups and spam) and the TiVo digital video recorder. Pop-ups and spam are direct extensions of the kinds of ads consumers were used to getting in the analog world, but with an added attempt to force consumers' attention. Pop-ups are magazine-like ads that thrust themselves onto a person's screen before or after a person views a web page. Spam, a direct descendent of junk mail, imposes itself into an internet user's email in-box without prior permission. The presence of both forms accelerated in the late 1990s as the internet bubble burst and caused a collapse of the online advertising market. Web media firms began offering potential advertisers opportunities beyond static ads in order to catch the attention of web users.

The growth was steep. Nielsen/NetRatings estimated that from the first to the second quarter of 2002 the number of pop-up ads grew from about 3.9 billion impressions to nearly 5 billion.[36] The tactic irritated large numbers of consumers and encouraged the creation of programs to block their appearance. In 2003, America Online announced that it would no longer allow pop-ups on its websites and would give its 34 million users software to block pop-up ads on other websites. AOL, once a champion of this form of advertising, changed its tune in a bid to hold on to subscribers who were in danger of leaving for internet service providers such as Earthlink that had less commercial clutter. (Earthlink had earlier begun promoting pop-up blocking as a feature of its service.[37])

AOL also trumpeted its presence at the forefront of trying to eliminate spam, which it defined as "unsolicited bulk email."[38] In early 2003, the company won a major $6.9 million judgment against a firm accused of sending more than a billion junk email messages promoting sexually explicit websites to AOL customers. A website announcement of its victory promised more court activity and added that it would improve its automatic spam filters. Jon Miller, the new chairman and CEO of AOL, said in a statement: "As a member, and as a parent, I too have become outraged by the tide of spam that's drowning the legitimate email I want to get. Spam is not only unwelcome on AOL, but we must make it unacceptable. We've declared spam to be public enemy number one on our service."[39]

AOL's corporate attitude reflected a concern with huge revenue losses, an executive shake-up, and a desire to stabilizing its membership. An AOL spokesperson said that customers identified spam as "their top concern" when using AOL, and competitors had tried to lure its customers by promising better spam filtering. As one industry analyst noted, "the companies that can help their customers deal with the increased level of frustration are the companies that can retain their customers."[40]

Business users of email also took up the mission of eliminating spam, because of the cost of carrying such messages, the time wasted opening them, and the uncomfortable environment that some of the ads caused for employees. And internet engineers and computer theoreticians were both annoyed and aghast that the world that they had worked so hard to build was being inundated by a commercial tidal wave. The statistician Paul Graham, a leader in the bid to destroy spam, defined the activity far more broadly than most commercial marketers would feel comfortable with:

I propose we define spam as unsolicited automated email. This definition thus includes some email that many legal definitions of spam don't. Legal definitions of spam, influenced presumably by lobbyists, tend to exclude mail sent by companies that have an "existing relationship" with the recipient. But buying something from a company, for example, does not imply that you have solicited ongoing email from them. If I order something from an online store, and they then send me a stream of spam, it's still spam.

For Graham, the goal was to create sophisticated filtering programs to put them into a junk file:

> To beat Bayesian filters, it would not be enough for spammers to make their emails unique or to stop using individual naughty words. They'd have to make their mails indistinguishable from your ordinary mail. And this I think would severely constrain them. Spam is mostly sales pitches, so unless your regular mail is all sales pitches, spams will inevitably have a different character. And the spammers would also, of course, have to change (and keep changing) their whole infrastructure, because otherwise the headers would look as bad to the Bayesian filters as ever, no matter what they did to the message body. I don't know enough about the infrastructure that spammers use to know how hard it would be to make the headers look innocent, but my guess is that it would be even harder than making the message look innocent.[41]

These were words, and activities, that worried many in the Direct Marketing Association, which until the late 1990s did not support legislation against unsolicited email. Not often spoken was advertisers' fear of what the most utopian of computer theoreticians were trying to bring about: the filtering of all automated email—"legitimate" or not—from people's in-boxes with such accuracy that people would not even have to look at those messages.

The same kind of worries also appeared closer to the center of national advertisers' world with the appearance of the digital video recorder (DVR), also called the personal video recorder (PVR). Essentially a computer with a large hard disk, the DVR acted like a VCR in enabling its owners to record programs and view them at other times. Unlike a VCR, the technology marketed to the public by TiVo and other firms was connected to an updatable guide that made finding programs across more than one hundred channels easy. Also unlike a VCR, in some versions made by Replay (and in "hacked" versions of TiVo) it allowed viewers to skip ahead 30 seconds at a time without at all viewing what was skipped. That, advertisers knew, would be commercials. In fact, Replay used its PVR's facility for skipping over commercials as a selling point in its early ads.

Both TiVo and Replay tried to assure advertisers and media firms that they weren't fundamentally out to destroy commercials. TiVo, in fact, invited advertisers to buy time on its service to download ads that might appeal to its DVR users. Those who said the fear of DVRs was exaggerated pointed out that sales figures were rather small. One trade article in 2000 called the DVR "a technology in search of a business model.[42] Others disagreed strongly. They pointed out that the sales rate of branded DVRs was increasing and that home satellite firms and cable systems were beginning to integrate unbranded versions into set-top boxes. They noted TiVo's admission that 60 percent to 70 percent of people watching via its technology were skipping commercials. And they admonished that whatever accommodation advertisers would make with DVR firms, it would undercut the by-then-traditional approach of mounting 15- or 30-second commercials within shows.

Jamie Kellner, CEO of Turner Broadcasting, warned that DVRs were destructive to the television business, contributed to lower ratings, lower ad revenue, and fewer quality programs for television distributors. "What drives our business," noted Kellner, "is people selling bulbs and vacuum cleaners in Salt Lake City. If you take even a small percentage away, you are going to push this business under profitability."[43] In describing the problem, Kellner made explicit the implicit contract between audiences and media firms about the need to attend to advertising. In 2002 he told the magazine *Cable World* that DVR users were "stealing" television by skipping the commercials. "Your contract with the network when you get the show is you're going to watch the spots. Otherwise you couldn't get the show on an ad-supported basis. Any time you skip a commercial or watch the button you're actually stealing the programming." When his interviewer asked him "What if you have to go to the bathroom or get up to get a Coke?" Kellner responded: "I guess there's a certain amount of tolerance for going to the bathroom. But if you formalize it and you create a device that skips certain second increments, you've got that only for one reason, unless you go to the bathroom for 30 seconds. They've done that just to make it easy for someone to skip a commercial."[44] "I am not against PVRs," Kellner said on another occasion. "I think it's an interesting technology. The only problem that I have is that the industry cannot continue to produce programs as it currently does unless it is either paid for viewing the programming, some kind of subscription model possibly, or people don't skip the commercials."[45] Another television executive agreed: "Somehow [original programming] has to be monetized to cause someone to make it. You do it through commercials, or you do it through some other form of payment, but don't expect it to get made for no reason whatsoever."[46] The verities of twentieth century advertising seemed to be crumbling. Despite actions by TiVo and other DVR makers to provide special areas for advertisements on their devices and to encourage viewers to go there, marketers' concerns about how they would reach their target audiences in the not-too-distant future escalated. In 2005, Yankelovich Partners, a market-research company, shared with media executives its finding that 69 percent of American consumers "said they were interested in ways to block, skip, or opt out of being exposed to advertising."[47] Fear continued to spread that rapidly diffusing technologies could make mulch of their traditional approaches to buying advertising.

To make matters worse, belief in ratings—those arbiters of twentieth-century advertising decision making—began to crumble. Scandals in which newspaper and magazine employees falsified circulation numbers made advertisers worry about how many people were really reading periodicals. Fundamental questions about the Nielsen television data raised the troubling question of whether young men had really abandoned traditional television for video games and the web or whether the data that suggested they had done so were wrong. Nielsen's discovery that the young men had returned did little to quell the suspicion among television and advertising executives that the ratings company was using outmoded methods to explore an audience it didn't understand in a territory it didn't know.

Clearly, the historical anxiety that media and advertising practitioners felt toward their circumstances had not changed in the century between the Truth in Advertising Codes and TiVo. The worry was that now consumers really could push away ads better than ever. In 2003, a Hollywood talent agent urged marketers to recognize that they could no longer present ads through home-based electronic media in traditional ways. "The genie is out of the bottle," he asserted.[48] The talent agent had a plan: product placement. Others had other plans, and they involved direct-response advertising. Neither was new, and neither was really a solution in its current form. As it turned out, though, both did point the way toward profound change.

3 :: Drawing on the Past

In September 2004, when many marketers complained that the just-concluded Advertising Week in New York had celebrated tired ideas, the TV talk show star Oprah Winfrey and the Pontiac division of General Motors did something that a lot of people in the advertising industry saw as electrifyingly new: Every one of the 276 people in Oprah's studio audience received a new Pontiac G6 sedan worth $28,000.[1]

The negative reactions to Advertising Week 2004 were startling for their criticism of Ronald McDonald, the Trix rabbit, the Energizer bunny, the M&Ms, and the Aflac duck, which characters had just been selected for inclusion in a Madison Avenue Walk of Fame. These were icons that had shaped Americans' relationships with advertising in the twentieth century. Ten years earlier, the time and creativity that advertising practitioners had spent to assimilate these characters into the popular imagination would have undoubtedly generated far more appreciation than a marketer's multi-million-dollar product placement on one daytime talk show. But this was a new era. *Advertising Age* editorialized that next year's Advertising Week should "home in on issues—marketing effectiveness, changes in technology and consumer behavior—that keep [its readers] awake at night."[2] The head of a major public-relations firm added: "It's nice to emotionally reconnect, but I don't think you should hark back to the good old days."[3]

Oddly, harking back was exactly what Oprah and Pontiac were doing—but in ways that were quite different from those celebrated by Advertising Week. Highlighting products in programming was a marketing tactic older than Bob Barker, host of *The Price Is Right*, a television game show that had been giving away cars (albeit one at a time) for decades. Yet here was a major trade paper calling the Pontiac-Oprah giveaway a public-relations victory. "I TiVoed it," the magazine quoted a marketing executive's major compliment. "It was so emotionally uplifting."[4]

Clearly, marketers' laudatory reactions to the Pontiac giveaway mirrored their views of elements from the past that would help advertisers in the future. What intrigued them about Oprah's angle was not only the give-away. It was that handing the audience cars was just the opening salvo in an imaginative integrated marketing scheme—a series of related activities aiming for buzz that would link the Pontiac to Oprah and her audience across a number of media. The giveaway show itself had a video of the host herself traveling to the GM plant in Orion, Michigan to inspect the G6 and certify its acceptability for her audience. That material was placed on the show's website. There was also a strong dose of web activity that included an opening for direct marketing.

One of Pontiac's aims was to drive potential customers to its website so that the firm's direct-marketing activities could go into gear. By offering a "Dream It. Win It" sweepstakes—the chance to win one of four "perfor-mance models"—Pontiac could get email and postal addresses that the firm could use to contact potential buyers directly. Whether Oprah's view-ers were likely prospects for the car is open for debate. What is clear, according to Pontiac, is that a link from the Winfrey site to Pontiac.com impelled 250,000 individuals to visit Pontiac.com on the day of the show, an all-time record for the site. The car maker added that within two weeks of the event it had achieved 87 percent awareness of the G6 among adults and the highest-ever Google.com click-through rate—that is, the largest percentage of people viewing an ad on Google who then clicked on it to go to the advertiser's website.[5]

Pontiac's marketing director cited the Oprah program's projections that the car giveaway would generate $20 million worth of unpaid media cov-erage and public relations.[6] Perhaps more important to marketing and media observers was that the integrated use of product placement and direct-response marketing is indicative of new directions in their business. "Product placement" refers to a marketer's insertion of merchandise into content presented as entertainment or news. "Direct response" aims to get the consumer to answer in such a way that he identifies himself to the marketer. Marketing strategists increasingly see these two businesses—businesses once sneered at by mainstream advertisers—as vehicles for implementing successful solutions to deep problems besetting their industry. They believe that, in a cluttered, ad-zapping world of computers, gaming consoles, and cell phones, product placement and direct response can communicate persuasively to fidgety consumers. This view returns direct response and product placement to their rather highly regarded positions before the middle of the twentieth century. "Direct" practitioners were acknowledged as the kings of marketing communication until the 1920s. The importance of product placement was recognized in movie-making and broadcasting through the 1950s. Both businesses served clients vigorously after their most celebrated periods. Yet the sense of their centrality to marketing communication declined among mainstream media and ad executives. It was restored in the 1980s and the 1990s, when image-based approaches didn't seem to offer solutions to many of the challenges of the digital environment.

Tracking the roller-coaster reputations of "direct" and "placement" helps to assess their legacies and the reasons advertising practitioners are turning to them at the start of the twenty-first century. A look back also highlights how marketers, media practitioners, and their critics used arguments about the social value of direct and placement to press their particular interests. That has often involved claiming that the benefits of product placement and direct marketing to consumers exceed those of traditional advertising.

::

In its late-nineteenth-century manifestations, direct response related to "mail-order" and "direct-mail" advertising. Because well-stocked stores didn't exist in some parts of the United States, entrepreneurs bought mailing lists or advertised in newspapers and magazines, offering to send catalogs. Ads also promoted individual products that the seller would discount if the consumer clipped the coupon and returned it with cash. These activities accelerated with the growth of postal routes throughout the country, the introduction of inexpensive first-class letters (the penny stamp in 1862), and the spread of rural free delivery routes in the 1890s.

Many of the offers mailed through the postal service were perfectly legitimate. Some of the most important and reliable businesses in the United States started as mail-order or direct-mail merchants. Tiffany, Orvis, Montgomery Ward, Sears, Burpee, and other firms reached out to America's many rural homes and found willing customers who could be encouraged by offers of free trials and instalment payments. As a history issued by the Direct Marketing Association notes, the catalogs of such companies "were appropriately referred to as the Farmer's Friend. Frequently, it was from them that farmers first learned of new mechanical equipment that would vastly increase their productivity. Their wives discovered sewing machines and other labor-saving devices that would reduce their workloads. Similarly, the seed nursery catalogs of the nineteenth century helped teach farmers which grains, fruits and vegetables to grow on the land they were settling."[7]

Nevertheless, the reputation of direct response suffered because the public often associated it with disreputable activities. The crusades of Anthony Comstock and his allies against objectionable reading material advertised and distributed through the mail established some of that reputation. Local merchants, fighting for their livelihood against what they called the Mail Order Trust, kept up a negative drumbeat, alleging undignified and extravagant claims by direct merchants.

It was the huge patent-medicine business that particularly gave direct response an image problem at the turn of the twentieth century. Very few of the products were actually patented. Not many of their creators would have willingly disclosed the ingredients—often a combination of herbs and roots with arsenic or alcohol. In a country with few physicians, at a time in which even the physicians who plied their trade knew little about curing diseases, and when women would much rather take tonics than undress in front of a doctor, these elixirs made psychological sense. Because social norms supported the products, they were advertised in even the most reputable media. As one advertising history notes, "so fastidious a magazine as the *Atlantic Monthly* carried in 1868 notices for Dr. J. W. Poland's 'Humor Doctor, an Invaluable Medicine for Purefying the Blood.' The *Atlantic* also happily sold a back-over half page to Turner's 'Tic Doloreaux or Universal Neuralgia Pill, the Undoubted Cure for All Excruciated Ills.'"[8]

Many patent-medicine firms made a lot of money from the sales generated by these ads, and they plowed much of the cash back into advertising to yield even greater revenues. The number of companies was so large, and the amount of advertising space they bought was so great, that in the three decades after the Civil War patent medicine was the main support of the great majority of American magazines and newspapers.[9] The contracts that big patent-medicine firms set before print media firms stated that the agreement would be voided if they carried articles detrimental to the companies' interests or if state or national lawmakers passed anti-patent-medicine legislation. Publishers and editors paid obedient attention, and the industry's influence on editorial matter was consequently immense.

In the 1890s, though, a few media outlets, most notably the *Ladies Home Journal*, began to buck the trend by refusing to carry ads for patent medicines and by publicizing the medical danger in the potions and the ways their advertisers were exerting power over news. The medical establishment joined the fray. By 1906, when Congress passed the first Pure Food and Drug Act, both the patent-medicine business and the advertising business had gotten public black eyes. Direct response was especially tainted. It was associated with fraudulent ballyhoo, in the eyes of the growing number of advertising practitioners who worked for goods that could be bought in retail outlets. They preferred advertising that aimed to sell a product by associating it with an entertaining message or creating a personality around it.

The Lackawanna Railroad's ads of the 1890s and the 1900s featuring a fictional passenger named Phoebe Snow provide an example of the new approach. In a series of more than fifty rhymes, the "Road of Anthracite" advertised its use of the clean-burning coal by telling the romantic saga of a pretty young woman who managed to keep herself attractive and her white dress clean despite a long time on the train.[10]

Through much of the twentieth century, the relative merits of direct-response and image advertising were debated. "Direct" practitioners argued that their business enforced a discipline that made selling the product paramount. Several of the best patent-medicine advertising writers moved over to write ads for more legitimate products using the same rhetorical techniques that had sold the nostrums. They had created a great many rules for selling and an enormous amount of lore about how to get fast results. By comparing the effects of different appeals, headlines, pictures, coupons, fonts, and other elements on sales, copy creators could learn quite quickly what worked and what didn't. They looked down on general announcements that didn't complete a sale, and they especially scorned image advertising. They argued that while it might be pretty or funny or cute, it could not bring the return on investment that direct response could with crafty ads that virtually compelled magazine and newspaper readers to send in coupons for products.

The practitioners of image advertising castigated what they claimed was a problem built into the rhetoric of direct-response advertising. They argued that in an era when consumers were constantly confronted with ads, sharply worded exhortations to buy would inure people to commercial announcements generally. Hard-sell announcements using some direct psychological and verbal tactics might be useful for the quick sale of inexpensive goods. Some practitioners of image advertising did borrow techniques from their direct counterparts in acknowledgement that ultimately even beautiful or humorous ads had to move products. At the same time, they insisted that the rhetoric of direct marketing could not develop brand identity, or personality, which was important to ingratiating products with customers over the long term.

By the 1930s, "direct" advertising had split off from the business of crafting announcements and images, and each trade looked down on the other. National consumer marketing clients sometimes used both, but they bought those services from different firms. The same was true when it came to other areas of marketing that involved direct response from the customer. Rarely did national advertising agencies get involved with creating and distributing newspaper coupons or help place attractive posters or setups at point-of-purchase locations such as supermarkets. Executives in the advertising establishment and its trade press dubbed these types of activities "below the line" work; the implication was they were not nearly as important or even as respectable as mainstream advertising. Aesthetic, ethical, and practical differences divided direct marketers and image practitioners for decades to come.

::

Like "direct," product placement was central to marketing for a long time only to lose out to straightforward advertising in the course of the twentieth century. The historian T. J. Jackson Lears points out that the insertion of commercial messages inside narratives is as old as marketing itself. Modern marketing evolved out of the carnivalesque atmosphere of the traveling markets of early modern Europe, from the 1500s to about 1800, where the selling message and the emotional engagement were intertwined in a form of public performance. "The market fair," Lears writes, "brought locally rooted townsfolk and peasants into contact with the exotic and the bizarre: with magicians and midgets, quacks and alchemists, transient musicians and acrobats; peddlers of soap from Turkey, needles from Spain and looking-glasses from Venice." It was, he points out, often hard to distinguish between the carnival and the selling. "Amid the carnivalesque confusion, market transactions leavened the imagery of abundance."[11]

Nineteenth-century American "medicine shows"—performances intended to sell patent medicine—were an example of the continuation of this blending of the theater and selling on the North American continent. A salesman named O. T. Oliver wrote of his attempts as "Nevada Ned" in the 1880s to sell a tonic called Hindoo Patalka by using "two Syrians out of a rum store and a Hindoo who was doing a magic act in variety, and [dressing them] in Oriental costume."[12] Such acts recalled P. T. Barnum's mixture of circus and product sales some years earlier. Lears notes that even relatively for "respectable advertisers" of the nineteenth century, the heritage of "theatrical exoticism" flowed into several forms of marketing— "print and pictorial, as well as sidewalk spectacles."[13] The seller often quite purposefully made it hard to tell the difference between the theater and the product promotion.

With the rise of professionalism in the newspaper business in the late nineteenth century, boundaries between news and advertising began to arise. Journalism codes taught reporters that good work meant being loyal to the "objective" ideals of reporting and resisting publicity and public-relations sources as well as pressure from the advertising department to support sponsors' interests. Along those lines, Henry Luce, the publisher of *Time*, proudly announced the separation in his organization of what he called the "Church" (the editorial process) from the "State" (the business domain). Insofar as Luce was both publisher and editor-in-chief, though, he violated the separation from the start. Moreover, reporters quickly learned that they often had to get along with publicists and PR flaks if they wanted to get good stories and exclusives. As the *Advertising Age* columnist Randall Rothenberg noted, "wily operators have crossed these borders for as long as they've existed. In the early days of mass media, the cleverest transgressors were the direct descendants of P. T. Barnum: the press agents who packaged stunts that could land their commercial clients in the newsreels."[14]

Entertainment companies got away with explicitly blurring distinctions between the commercial and the "purely" theatrical more easily than news firms. From the earliest days of Hollywood movies, in the 1910s, the studios were quite directly involved in placing products into their theatrical creations. For many years, the major decision to put a particular product in a film lay with the property master. If the script or the set design called for a car or phonograph, the property master might borrow one from the manufacturer to save the studio money. The firms soon inferred that audiences responded to such appearances by purchasing the products. Marketers consequently approached the property masters, sometimes with cash under the table, to offer their products as props. Studio executives began to realize that there was even more to be gained here. The appearance of a product in a film might reflect a deal in which the manufacturer paid the studio for placement and permission to have one of the studio's actors endorse a product, or the payment might be for placement and an agreement to promote the movie in magazine and billboard ads for the product.[15]

Studio executives and creative personnel argued that showing brand-name products enhanced the audience's sense of realism. Independent theater owners—those not owned by the major studios—didn't buy that reasoning. They were annoyed that the studios were not sharing the money from product placement and related promotions. Their discontentment was exacerbated in the 1930s by the studios' refusal to allow them to show filmed advertisements before the studios' movies.[16]

The major Hollywood studios had strong practical reasons for keeping ads out and placements in. They themselves for a short time created these sorts of advertisements—which they called "minute movies"—for theatrical release. Studio executives began to worry, though, that major advertisers might get the idea of funding independent film studios to create full-length movies under their sponsorship, much as they were sponsoring radio programs.[17] The major studios also heard from newspaper executives who did not want advertising competition from the movies.[18] Wanting neither to encourage advertisers to support movies nor to alienate newspaper executives, on whom they relied on for film publicity, studio executives acted in concert to refuse to allow commercials to be screened in theaters projecting their movies. They justified their action publicly by making ethical claims relating to their audiences. Nicholas Schenk, who ran MGM and its Loew's theater chain, said he was against the commercialization of the cinema "because it is unfair to our audiences. An advertisement on the screen forces itself upon the spectator. He cannot escape it, yet he had paid his admittance price for entertainment alone." Schenk added that the temptation was there to make advertising more and more obtrusive and so more and more annoying to the audience.[19]

Angry that a funding stream was being closed to them, the independent theaters lashed out at product placement publicly, with their own ethical claim about its effect on their customers. Concealed advertising, the exhibitors declared, was "obnoxious." They argued that audiences expected a program of 100 percent entertainment and should not, even unwittingly, be subjected to movie features containing paid advertising matter.[20]

In the late 1940s, the U.S. Justice Department forced the major studios that owned theater chains to divest them. Now the movie-makers could not rely on their own outlets to reach the public and had to be more sensitive to exhibitors' concerns than in the past. Perhaps it was this turn of events that led the studios to play down product placement. Although the activity continued after World War II, it was done quietly, often surreptitiously.

Product placement in radio paralleled the rise and decline of the activity in the pre-World War II movie industry. Placement developed in radio in part because of a concern by early stations that their audiences would feel that audio advertisements coming into their homes violated their privilege not to be bothered there. The notion related to late-nineteenth-century and early-twentieth-century notions of the sanctity of the home as a place of refuge, a "haven in a heartless world" of business and industry. In 1890, Samuel Warren and Louis Brandeis, tying into this belief, had argued in the *Harvard Law Review* that Americans had a right to be "let alone."[21]

In the early 1920s, when radios began to appear in American homes, such intrusion was not an issue; stations were owned and sustained by manufacturers of sets or stores that sold them, and there were no commercial messages beyond mentions of the firms' names. To Secretary of Commerce Herbert Hoover and to others, the idea that audio ads would flood into the home via the new medium was anathema. "I believe that the quickest way to kill broadcasting would be to use it for direct advertising," Hoover warned. "The reader of the newspaper has an option whether he will read an ad or not, but if a speech by the President is to be used as the meat in a sandwich of two patent medicine advertisements there will be no radio left."[22]

Even some members of the advertising industry would have banned ads altogether as violating the audience's right to be left alone. In 1922, for example, the trade magazine *Printer's Ink* argued that "the family circle is not a public place, and advertising has no business intruding there unless it is invited." The same year, *Radio Broadcast* urged its readers to write letters to their congressmen protesting the use of radio for advertising.[23] Their warnings went unheeded by AT&T and other station owners looking for new ways to profit from the ether, and within the next half-decade they had made sponsorship the norm. Moreover, within the next six years the notion that there were other ways to support broadcasting (for example, a tax on radio, or public support of educational stations) was swept away by broadcasters' political and economic power. In the five years before the Communication Act of 1934, the government allocated frequencies to organizations wanting to broadcast. During those years, commercial broadcasters pressured the Federal Radio Commission to make sure that stations like theirs received preference for frequencies over noncommercial educational and labor-union stations. They announced a clear ethical justification for advertising: Audiences were getting free programming, and it was only natural for them to support this material by allowing commercials into their home.

In the first few years of radio advertising—the early and mid 1920s—station rules confined commercial announcements to short, decorous statements, like some heard today on National Public Radio. The push to get around these rules was strong, however, and quite soon sponsors gave the stars of the programs promotional names—for example, The A&P Gypsies—so that mentions of them throughout the program reminded viewers who was paying for the show. It was a crafty combination of product placement and primitive advertising. By the 1930s, frantic Depression-fueled commercial competition by local stations and networks for many radio advertisers led to the abandonment of much of the decorum for hard sell and jingles, but they continued the odd combination of product placement and straightforward advertising. The advertising agencies that produced the programs for sponsors saw to it that writers inserted plugs into programs and built comedy routines around them (for example, Abbott & Costello's bit from the 1940s about whether it "Hertz" to rent a car).

Executives justified the practice by insisting that audiences liked this approach more than spots interrupting the action. Of course, audience typically got both by then. Still, it is an important corrective to typical tales of the relentless growth of advertising to recognize that when it came to radio of the 1930s and beyond, marketers and talent often championed not-so-subtle product placements over direct advertising. The combined approach also transferred to early television during its live-broadcast years of the late 1940s and the 1950s. A 1951 *Variety* article noted that "pluggers trip over each other" to get on the air. It said that virtually every public-relations agency assigned people specifically to ensure mention of a client's product. Producers, talent, or writers received from $75 to $125 for each plug. In one case, a comedian doing a guest shot on a television program worked in half a dozen plugs, "which earned him more than his performing fee."[24]

During the 1950s, certain forms of product placement in radio and television came under duress. In radio, government regulators fixated on evidence that record companies were paying radio personalities to insert specific records onto their playlists; the practice was called "payola." Federal lawmakers decreed that it was an ethically unacceptable violation of audience confidence to do that without telling listeners. In television, the sponsors that controlled a number of high-money prime-time quiz shows admitted that they had given certain contestants the answers in order to keep them on the programs and eject others. Federal regulators judged this practice, too, to be fraudulent.[25]

Although the quiz scandals didn't have anything directly to do with product placement, the angry government hearings that ensued along with loud Congressional anger around 1960 about television violence led television network officials to redefine their relationship to programming. Previously the networks typically sold air time to the advertisers; they and their advertising agencies then created programming they found appropriate. Increasingly in the 1960s the networks chose the shows to air and sold time slots in breaks within and between them to one or more advertisers.[26]

With new-found control over their schedules, network executives worried that product placements would not suit their interests. They fretted that mentions within programs that weren't revealed to the audience would go against new federal anti-payola regulations requiring broadcasters to reveal who was paying for what. Such disclosures would make sense on a game show but perhaps not on more prestigious evening ("prime-time") programming. Network officials also worried that product place-ment would contravene voluntary limits on advertising minutes. They had agreed to insert these limits into the National Association of Broadcasters' Code of Good Practice in response to militancy by advocacy groups and government concerns about too many television ads. Allowing products to be placed into programs along with free-standing ads would invite the kind of scrutiny by Congress that the television networks did not want. The result was that from the 1960s through the 1970s product placement in network programming was rare outside of quiz and game shows. One of the few exceptions was the placement of automobiles. The networks allowed car companies to donate the use of their models to producers for use by the characters on programs. The reason was that it saved a lot of money. The obvious justification was that hiding a car's brand identity from audiences would be difficult and possibly silly. A result was that various Fords, matched to characters, managed to populate many of the prime-time programs on all three networks. Apart from the cars, though, television's protagonists drank generic soda and ate generic cereal. It was a commercial-free world that was bracketed almost completely by brand-image commercials.

With the decline of product placement, image advertising emerged in the television era as the most "legitimate" form of marketing communica-tion among marketers in print, outdoor, and broadcast advertising. Direct marketing and product placement didn't disappear. The direct mail busi-ness became an important industry unto itself and a financial mainstay for the U.S. Postal Service. Product placement persisted in the shadows, often lumped with promotional marketing and public relations among "below the line" activities that were carried out by organizations ancillary to the real work of ad agencies.

To many in the mainstream advertising industry and media, image advertising was what counted—what won prestigious awards, what the public remembered and talked about, and what regulators didn't mind. In the 1960s and the early 1970s, there was luxurious creativity in advertis-ing. Advertising practitioners looked at magazine ads and television and radio commercials as works of art in the service of commerce. The imagi-native use of pictures, words, and (where possible) sound fit perfectly with the needs of media practitioners. It was a period of strong anti-commercial impulses. Image advertising, while still hawking products, didn't seem as sleazy as the insistent messages that network and magazine executives associated with direct selling. Many mainstream media executives, particularly those with journalism in their backgrounds, also considered image ads to be more honest than product placements, which smacked of breaking Henry Luce's "Church-State" divide by presenting paid-for messages as "genuine" editorial materials.

::

Developments that began in the 1980s once again put direct response and product placement on a trajectory to vie with image advertising for marketing's central stage. The change was related to marketers' need to use targeting to confront the growth of cable, VCRs, and computers that led an unprecedented fragmentation of audiovisual channels to the home and even out-of-home, in supermarkets and malls—the technologies of division discussed in chapter 2. It was a time when competitors' claims were cacophonously bumping up against one another in a clutter of advertising vehicles that ranged from television to billboards to race cars.

Two movie placement bonanzas—one in *ET: The Extra-Terrestrial* and one in *Risky Business*—got marketers and production firms streaming toward product placement as a way to addresses these problems. The ET placement story is legendary: The producers contacted the Mars company to ask permission to use M&Ms in their film as the candy that cemented a friendship between a young boy and a creature from another planet. Mars turned the idea down, so the producers used Reese's Pieces without the knowledge of the manufacturer (Hershey). While the movie was being made, the producers asked Hershey to put money into a tie-in promotion of the film and the candy. Hershey agreed to put up $1 million.[27] The results astounded marketers: Sales of the product reportedly jumped 65 percent in the first three months after the movie's release. The apparent power of product placement showed up again that year when Tom Cruise wore Ray-Ban Wayfarer sunglasses—a model which Ray-Ban executives feared was on an unstoppable decline—in the film *Risky Business*. After the movie's debut, Wayfarer sales allegedly increased spectacularly.[28]

It was claimed that inserting products into films did well what traditional advertising could not often do at all: rivet a hard-to-reach target audience's attention on a product at a time when agencies' frenetic attempts to reach various segments of the population had created ad clutter throughout the media landscape. Movie theaters provided a quiet place where captive audiences could see an exclusive product perform in a realistic environment. That, at any rate, was the argument. A seminar titled "How to Market Your Product in Motion Pictures and Turn the Silver Screen into Gold" attracted several dozen executives. Placement on an organized basis being rather new, only a few placement agencies then existed. The companies not only promised to put clients' products in films; they promised to make sure that the products weren't included in films or in particular scenes that might make the brand look bad. The biggest placement firm of the early 1980s was Associated Film Promotions (AFP), which had been founded in the late 1970s. According to Janet Maslin, a film critic for the *New York Times*, AFP charged companies fees starting at $35,000 a year for handling their products. In exchange, AFP guaranteed that the product would make at least five movie appearances. Maslin asserted that money did not "usually change hands" with the studios, though the cast and the crew members might get some gifts. AFP's head, Robert Kovoloff, said: "We're like marriage brokers; we save film companies time and money."[29]

Despite Kovoloff's aim to make his practice seem like a traditional exchange of favors, it was becoming clear that there was sometimes a lot of cash available to studios interested in working with marketers. As 1983 ended, Twentieth Century Fox became the first studio to publicly offer manufacturers a specific display of their brand-name products in movies in return for cash payments of $10,000–$40,000. The studio offered no guarantee that the director would not in the end edit the scene out of the movie, so the manufacturer did have the right to get the money back. Nevertheless, the announcement encouraged more companies into the fray and ratcheted up the entire process of negotiating the nature and prominence of placement.[30] A professor of business at the University of Southern California further promoted the activity's value for convincing audiences. "This is a form of advertising that you simply can't buy elsewhere," said Ben Enis. "When advertising is labeled as advertising . . . your guard is up. But in the movies you can have the hero or heroine implicitly or even explicitly endorse the product. It's quite effective."[31]

By the late 1980s, product placement had become an integral part of the marketing process. A company called CinemaScore arose to measure the percentage of theater audiences that recalled particular inserted products.

Agency holding companies such as Young & Rubicam began to buy placement firms; an executive from one of these contended that "the vast majority of Fortune 500 companies are involved in getting their products placed."[32] A major reason for the commercial rush into movies by the late 1980s had to do with marketing executives' awareness that the fractionalizing media environment meant that theatrical movies would be seen far beyond theaters—most prominently, on home VCRs and television channels. Placement executives extrapolated that movies making certain sums at the box office would sell certain numbers of VCR tapes and achieve certain ratings when broadcast on network television, then on cable television, then on local television.

Though there was always a risk that the movie might flop , the number of eyeballs a hit could reach moved the business forward. For example, in 1990 Pepsi-Cola paid millions of dollars for a "promotional partnership" with Twentieth Century Fox to help advertise the film and videocassette versions of *Home Alone* in return for the appearance of its flagship brand in the movie. The film turned out to be a monster hit, and Pepsi was delighted. The movie relationship proved fleeting, however. Fox forced a bidding war between Pepsi and arch-rival Coca-Cola for placement rights in *Home Alone 2*, which Coca-Cola won.[33]

The interest in product placement as a vehicle for showing off products in a cluttered media environment was strong, and it was only a matter of time before it was reborn in prime-time television. The movie industry had cleared the path. In the Reagan era, the Federal Communications Commission no longer seemed obsessed to jawbone down the number of commercial minutes the networks aired in prime time, preferring to let competition with cable set the standard. In the 1980s, first-run syndicators—companies that produced new series for local stations rather than networks—began to use product insertions as a way to cover costs.[34] So did cable television networks. In fact, two of the earliest cable networks, Music Television and the Home Shopping Network, might be said to have been all placements all the time. Founded in 1981, MTV was built around videos sent free to promote recordings. The Home Shopping Network brought some of the carnival aspect back to selling products.

Other cablers were offering product placements in programs as added values when they bought regular commercial time. One consultant saw meeting advertisers' demands in this way as a by-product of cable's need to compete with the broadcast television networks, which had not been so open to such deals. "The answer is for cable to make sure it is responsive to marketing needs vs. simply providing eyes," he noted.[35]

In fact, broadcasters were beginning to feel the competition from cable by 1988. By then, more than half of Americans had cable television, and the three major broadcast networks' shares of the prime-time audience were beginning to dip noticeably. A related problem for marketers at the time was the increased use of the remote control to switch among the new cornucopia of channels. A 1988 article in *Advertising Age* remarked that the generation that grew up with the remote control had reached adulthood, "bringing with it limited attention spans and itchy remote-control trigger fingers." The article reported research that claimed more than 50 percent of viewers aged 18 to 34 were watching more than one program in a half hour period, with 20 percent watching three or more." These and other findings led advertising executives to worry that people were not staying around for their commercials. Advertisers had tried to make their commercials shorter and more entertaining to hold viewers, but that didn't seem to always work. Consequently, by 1990 some advertisers were attempting to mute the effect of commercial zapping by making deals with program producers for product placements on network shows. Advertising-agency executives predicted, however, that network executives would clamp down on the practice.[36] The major broadcast networks remained hesitant about the practice, at least partly because of a tempest in the movie industry that reverberated on them.[37] As film producers revved up their placement activities, the Center for Science in the Public Interest began to make noise about a loophole in the FCC rule that allowed movies airing on TV to brandish products, including cigarettes, without informing audiences that they were paid insertions. The CSPI argued that the practice was merely another form of payola, and that it was "important that people know when advertising material is appearing in broadcast movies."[38] To make problems worse, U.S. Representative Tom Luken, a Democrat from Ohio, proposed that the practice was paid advertising. A bit later, at the request of CSPI, the Federal Trade Commission began to investigate film product placement, especially by cigarette makers.

The negative news revived discussion among the press and regulators of the ethics of product placement. The bad publicity seems to have encouraged the producer of the James Bond movie *License to Kill* to place a warning against smoking in the closing credits. Kerry Seagrave argues in a chronicle of movie product placement that apart from that warning "all the furor raised by Luken and CSPI produced no results."[39] Yet the tempest did energize the business to control its reputation. Product placement firms started a trade group, the Entertainment Resources & Marketing Association. *Advertising Age* wrote acidly at the time that "the association's primary goal seems to be legitimizing the now somewhat fly-by-night image of its business."[40] Later renamed the Entertainment Marketing Society, its members and other Hollywood production firms made much about their insertion of messages about AIDS and other diseases into movies and television programs. They made friends with health foundations and tried to get the press to associate the movie and television insertion business with good intentions as well as commercialism.

By 1993 the broadcast television networks had decided that prime-time product placement was no longer a liability. That year, NBC, ABC, and Fox proudly publicized that they would allow live or taped commercials featuring regular series cast members to appear during the programs. It was also the year that NBC's new hit comedy show *Seinfeld* began showing characters using actual products. Jerry Seinfeld, the show's main writer and its star, suggested that it would enhance the audience's sense of realism about the show. "We like to have real products in the show," Seinfeld told *Advertising Age*, insisting that none was a paid insertion. The only problem seemed to be that paying advertisers for the show objected. "To create a realistic set without offending its advertisers," the trade magazine noted, "the Seinfeld crew tries to use a variety of brands on camera."[41]

The *Seinfeld* production team may well have been genuine about its desire to use branded products for verisimilitude; the mundane names provided an odd counterpoint to the cast's sometimes surrealistic situations. Still, reality as a vehicle for audience identification or satisfaction had been offered as an excuse for product placement in the movies going back to the 1930s. For 20 years in which generic products were generally used in prime-time programs, though, no critics or audience groups seem to have raised the policy as a drawback to enjoyment. In fact, when critics did bemoan a lack of realism in prime-time programming, their concerns never seem to have revolved around the need for name brands.

The real momentum building for product insertions in prime-time broadcast television shows, of course, was that making money from the activity was increasingly practical at a time when drawing profits out of broadcast network television was getting harder. In an environment where viewers had dozens of channels, where the cost of making a show might outstrip the ability of the producer to profit from it through only the sale of commercials, and where international sales and reruns were iffy, product placement suddenly seemed like a logical step. ABC, CBS, and NBC initially may have felt left out of the process by Federal Communications Commission rules, adopted in 1970, that had effectively prevented them from owning or leasing reruns (that is syndicating) of most the prime-time shows they aired. In 1995, the FCC eliminated the rules. The networks again started owning prime-time entertainment shows, and they quietly began to invite marketers to use product placement on a regular basis.

::

By the late 1980s, product placement was positioned as deserving renewed attention for its ability to help marketers at least as well as general advertising. Direct response had achieved that status about ten years earlier. That did not mean that direct response would move quickly into network television. Instead, it became linked to new forms of direct marketing, an established business focused on gathering names of individuals and targeting them by mail or phone. Overall, advertising practitioners were giving renewed respect to the emphasis of direct response on actually getting people to buy things.

The revived interest in direct marketing's rhetorical approach stemmed in part from the difficult economy of the late 1970s and the early 1980s. Soaring oil prices and interest rates, heightened Middle East tensions, and a growing sense that Japan was surpassing the United States in innovation and efficiency led to a sluggish selling environment. Some mainstream advertising practitioners wondered if their traditional image-oriented toolkit was appropriate for the times. A January 1982 editorial in the newsletter *Ad Day* captured the changing mood nicely. It began with the proposition that "these are not the best of times" and "they are not likely to get better by next week or next month." The writer then opined that "it is certainly no time for an advertiser to shell out honest dollars for ads and commercials that simply decorate the scenery or amuse and titillate bored and slack-jawed audiences."[42] The writer continued acidly: "Yet how many advertisements of TV commercials do you see that move and shake people—stir 'em up—crank 'em up—ask them or jolt them to get of their duffs and do something. Damn few. Most of them don't even try. They seem smugly satisfied to simply spray a little information around—exude a nice happy glow—or leave an inoffensive impression. Beyond that is considered hard sell—and hard sell is non-chic these days."[43] The editorial went on to contend that the times required "ask for action" advertising, and that direct-response advertisers could provide a model. Practitioners of direct marketing knew how to select audiences in ways that mainstream advertisers, with their traditional focus on the mass market, had not learned. They knew "how to develop a sales argument—how to clinch a sale." They knew how to really talk to an audience so as to "offer a proposition the reader finds hard to refuse." "No mass audiences here— no prime-time millions of faceless people. The direct-response advertisers build lists from carefully screened names and addresses—and once they convert a name into a buying prospect they hold on to them year in and year out selling them items. Mass advertisers obviously can't do the same— but there is a lesson here: customers come one at a time. They are sold person by person. It's worth remembering."[44]

In 1982 the idea of using television to sell to one customer at a time seemed impractical, not to say silly. As the *Ad Day* editorial noted, when it came to customer lists direct response operated mainly in the print domain, though telephone marketing was growing. Nevertheless, the past decade had seen a revolution in direct-response advertising. Many observers of the marketing business were already prognosticating that it represented the future of selling. Changes in technologies converged with changes in marketers' mindsets to establish the radical idea that mainstream advertising practitioners needed to reach out to their audiences using hard sell techniques that they believed in past years were distasteful, almost unethical. Now updated versions of those techniques were repositioned as ways to communicate that audiences would find efficient and relevant.

The technology that led direct marketers to revolutionize their industry was the computer. It encouraged the rethinking of both the nature of the audience and ways to access them. Direct mailers had used lists of potential customers in the nineteenth century.[45] The computer made it easy to store, combine, and cross-tabulate many different lists. Large direct-marketing firms began to keep names, addresses, and other information on computers in the 1960s. In the 1970s, consultancies emerged to take advantage of the computer's ability to merge large databases for marketing purposes and perform new kinds of number-crunching analyses on the newly merged files. Two such companies, SRI International and Claritas Corporation were presenting direct-response firms with new ways to seg-ment the American population for efficient selling. Claritas' PRIZM offer-ing, for example, used cluster analysis on the U.S. Census and databases from other firms to classify every ZIP code into 40 lifestyle groups. The idea was to categorize neighborhoods by demographics as well as the kinds of media habits, possessions, and purchasing opportunities they would likely have, "right down to the cereal in the cupboard and the antacid in the medicine cabinet."[46]

The computer's ability to store and sort the names of millions of people and their characteristics moved the list business—now called the database business—into overdrive. The buying and selling of names and informa-tion about them became a major industry. Broadly speaking, direct mar-keters used two resources for name gathering: universal databases and transactional databases. Universal databases are compilations of informa-tion on every individual and household in an area, even an entire nation. Inferring buying interest from a range of demographic, psychographic and broad lifestyle information was often helpful as a starting point. Yet many direct-marketing practitioners of the 1980s and the 1990s followed the long-held dictum of their business that led in a different direction: the best predictor of future behavior is specific past behavior. This proposition underscored the importance of transactional databases. A transactional database is a list of people who explicitly responded to a particular market-ing or fund-raising appeal. Marketers would often purchase names from other marketers with products that reflected compatible lifestyles. For example, names of men who paid for season tickets to sporting events might attract a company trying to sell sports memorabilia. Similarly, a frequent-flyer list would likely draw a hotel chain with an eye on the busi-ness traveler. In the 1970s and the early 1980s, these segmenting tools were pressed into action. The 800 number encouraged quick responses to televi-sion and catalog offers, ink-jet printing enabled mailers to personalize mes-sages to individuals, and the personal computer allowed easy storage of data and access to sales results. It was a boom in innovations that allowed for creating and reaching social segments at a time when respected market research such as Yankelovich Monitor, the marketing trade press, and popular pundits were saying that American society was more frenetic, divided, self-indulgent, and suspicious than ever. The converging developments generated a conviction among direct-marketing practitioners that the advertising world was moving their way. A Dun and Bradstreet executive saw the social fragmentation as paralleling the multiplication of media channels and the distribution of audiences across more outlets than ever. The traditional hallmarks of direct marketing—precise audience identification, individualized media communication, and speedy, full-satisfaction order fulfillment—could mesh constructively, he said, with the new technologies.[47] The futurist John Naisbitt was even more blunt. "Direct marketers are at the forefront of where everybody is going to be," he predicted in 1983. "We can all learn from them."[48]

From the late 1980s on, the amount of direct marketing was unprecedented. The number of television commercials that invited viewers to use 800 numbers and credit cards to buy products by mail increased dramatically, especially on cable. Cable and independent outlets were also vehicles for long-form ads called "infomercials." Taking the informercial one step further were shopping channels that invited immediate purchases by phone 24 hours a day. Still, while television-based platforms for consumers' responses were especially visible in the 1980s and the early 1990s, it was via the mail and the telephone that most targeted advertising took place.[49] Catalog mailing and other solicitations also increased through the decade. The numbers were dramatic. In the second half of the 1970s, advertisers doubled their direct-mail expenditures; they reached $10.5 billion in 1981.[50] According to *Direct Marketing*, mail expenditures soared from $12.7 billion in 1983 to $17.2 billion in 1986. By 1989 direct-mail spending had climbed to $23.4 billion, and in 1993 it was $27.3 billion.[51] Use of the telephone for marketing, often to order from the mailed catalogs, grew even more dramatically. According to *Direct Marketing*, companies spent $34 billion in 1984 on marketing transactions via the telephone. By 1990 that number had risen to $60.5 billion, and in 1993 it was $73 billion.[52]

Traditional advertising agencies were beginning to see the handwriting on the wall. By 1982, fifteen of the top twenty advertising agencies had bought or started a direct-marketing capability.[53] It was primarily their clients' growing interest in computer-guided targeting of niche markets that brought them to look at direct-marketing practitioners with grudging respect. Philosophical and aesthetic splits between "direct" and "image" practitioners still existed, but having the two groups under the same corporate umbrellas perhaps opened possibilities that they could learn from one another.

Direct response shared a number of characteristics with mainstream advertising. Direct marketers' penchant for dividing consumers fit with the movement toward increased specificity about audiences in those other activities. Nielsen television audience data; MRI and Simmons syndicated data on demographics, psychographics and purchasing patterns; geodemographic extrapolations from database firms—these and other storehouses of information were scavenged by creators of image advertising as well as by tacticians of direct marketing. And in "direct" work, as in "image" advertising, the focus was on reaching out—signaling—to a particular population with certain categories.

Where the direct-response business diverged from the mainstream, and where marketers saw its greatest possibilities, was in the connections that its practitioners tried to make with their targets. The 800 number and the personal computer made it easier than in the past to link up with potential customers quickly. The technologies also made it easier than before to track the "pull" of an ad.[54] On a cost-per-thousand basis, this approach was clearly more expensive than using magazines or network television. Direct-marketing practitioners insisted that what they lost in efficiency of reach would be more than made up in the careful selection of people likely to act on the sales pitch. The trick was to get good lists of likely prospects.

In the 1990s, however, more and more direct marketers began to believe that, as important as prospecting for new customers was, they should pay more attention to the customers they already had. The reason was the finding that a high percentage of a company's profit comes from repeat purchasers and that it costs several times more to get a new customer as it does to retain a loyal one. It was actually the renewal of an old idea. The nineteenth-century Italian economist Vilfredo Pareto had proposed that 80 percent of his country's wealth was held by about 20 percent of the population. In the 1930s and the 1940s, the quality-management expert Joseph Juran elaborated Pareto's insight into a universal generalization he called the "vital few and trivial many" and inaccurately called it Pareto's

Principle.[55] The idea was used to describe phenomena across various industries.[56] For merchants, twentieth-century data seemed to confirm that 20 percent of a merchant's customers generated 80 percent of sales. Marketers regarded this "80-20 rule" as crucial to targeting the right audiences in an age of fractionalized media and highly differentiated, frazzled consumers.

Direct marketers put out the word that their ability to help a firm not only target and signal the right kind of customers but keep them was their greatest asset. Keeping customers, they said, requires establishing "dialogues" with the firm's consumers.[57] The aim was to reinforce repeat purchasing with signs that the company was tailoring its activities to their needs and those of the people like them.

The key to believing that it was worth the effort lay in recognizing a repeat customer's value over time, what some called the "lifetime value." It was clearly evident to airlines, whose frequent flyer programs (beginning in 1981 with American Airlines) were emblematic of the new approach to gain "loyalty."[58] Similarly, upscale store and hotel chains saw the utility of keeping updates about their customers and contacting them on regular bases. Direct-marketing practitioners called the practice "one-to-one marketing," or more commonly "relationship marketing," and they saw it as the new database-driven incarnation of their business. As the 1990s progressed, retailers and manufacturers of inexpensive "package goods"—diapers, cereals, soups, inexpensive cosmetics, over-the-counter pharmaceuticals—also moved toward tracking and wooing individuals one-on-one.

Intense competition for shelf space in supermarkets and department stores forced executives from even the largest manufacturers to find ways to explore the hypothesis that repeat customers, properly handled, could help keep brand prices up (because they would pay more for products that paid attention to them) and niche brands on the shelves. If a company saw a person's purchase of a box of corn flakes as a single incident, then direct marketing of any sort made no sense. Shooting a message at millions of people on network television made for a much lower cost per thousand people, even if the great percentage viewing the commercial did not end up buying the cereal. If, in contrast, the marketer could see that a known purchaser of corn flakes would be making one of thousands of decisions to purchase the corn flakes over a lifetime, then the consumer's cereal-buying habits would take on an entirely different kind of value. Taken further, if the marketer would identify the loyal consumer of corn flakes as someone who would likely buy a range of the firm's products over many years, the lifetime value of that person would be even greater.

The proposition presupposed getting information about those heavy users, and in the early 1990s marketers were increasingly adopting a variety of methods to gather data about their best customers. They were asking for information from consumers by means of sweepstakes forms, mail-in rebates, product-registration cards, in-box coupons, 800 numbers, and events aimed at their target audiences. Procter & Gamble's Metamucil bulk laxative, for example, was capturing the names of heavy users by means of 800 numbers and address forms on coupons. P&G was then periodically sending them detailed product updates and information, coupons, and samples of new products.[59]

Observers of the day considered such activities "cutting-edge." One consultant found that the number of firms collecting consumers' names and addresses had tripled between the end of 1991 and October of 1993.[60] Don Schultz, a professor of Integrated Marketing Communications at Northwestern University, saw it as a sea change for direct marketing. "Traditional direct marketing is nothing more than mass marketing with a response device," he noted in 1995. "It's marketing on the averages, where today we are moving rapidly into marketing on the differences. With databases, you can learn those differences and adjust promotion and advertising accordingly. But no matter how you promote and advertise, if you don't have a database soon, you'll be out of business."[61]

In 1993, direct marketers could sense the tide shifting toward them. "There's no question direct marketing has grown faster" than general advertising, said Jerome Pickholz, chairman of Ogilvy & Mather Direct. That firm was a subsidiary of the traditional advertising agency Ogilvy & Mather, which itself was a part of the huge WPP marketing communication holding company. "We're not telling clients that they should put all their money into direct marketing," Pickholz noted, "but it certainly makes sense to allocate some media spending in a way that reaches very specific target households." He added: "There's something wrong in an advertiser's plan if loyal users of a given brand get the same kind of advertising exposure as those that are not. We need to start getting across some special messages to heavy users, to treat them differently, because right now, we're treating them like the rest of the mass audience."[62]

It was a new approach to databases, explicitly noting an interest in favoring some consumers over others. "Loyalty marketers" insisted that separating the profitable 20 percent from the unprofitable 80 percent required it. Randy Petersen, a loyalty marketing expert, made the point bluntly to *Advertising Age*: "We argue strenuously, strenuously against naive sentimentalism on the part of companies who insist, 'We love all our customers and we love all our customers the same.'"[63] But in 1995, when Petersen made that comment, the technologies for gathering individual names were cumbersome. In addition, the vehicles for continual contact with customers were still rather traditional and expensive—direct mail, telephone marketing, fax—so that the costs per person were high, even if they were coming down. A study commissioned by the U.S. Direct Marketing Association found that "many retailers lack a complete understanding of database marketing" even though two-thirds said they had such programs.[64] The report concluded that retailers faced many challenges when it came to database marketing, a point echoed about consumer package goods by the consultant Robert Wientzen, who would later head the USDMA. Despite his enthusiasm for database-driven relationship marketing, Wientzen acknowledged that marketers were still looking for the best ways to maximize consumer response through these programs. "Package-goods marketers still are enthused about the possibilities of database marketing," he said, "but no one has yet cracked the code."[65]

Increasingly, people throughout marketing recognized the importance of cracking that code. A 1995 *Advertising Age* editorial noted the shift in the status of traditional brand-image advertising compared to direct response—that is, compared to work designed to "stimulate a direct order or a qualified lead, or to drive store traffic."[66] The editorial cited a study funded by the USDMA which found $12 of every $100 spent on consumer goods and services was being generated by direct marketing efforts. *Advertising Age* did not disagree with the study's prediction that "direct response techniques will increase their share of total consumer and business-to-business sales slowly but steady through the end of the 1990s." To the contrary, the editorial advanced the idea that direct work would inevitably grow in its power: "There will always be the need for great advertising that gives brands and companies an image and a personality. But managers' thirst for two-way contact with customers and potential customers is hard to quench. The adperson who is master of this particular form of "conversation" can expect a growing role in tomorrow's marketing world."[67]

::

Like product placement, then, direct marketing had moved closer to the mainstream of marketing in the mid 1990s than it had been in nearly 40 years. Like product placement, direct-marketing practitioners now could get away with arguing that they were more appropriate for audiences than the image advertising that had become traditional in the second half of the twentieth century. Marketers saw each set of activities as a solution to the technologies of division besetting their attempts to reach audiences. They did not see product placement and direct response as at all related, however; nor did they yet see them as activities that would really equal to traditional advertising work in visibility. That was about to change. The 1995 *Advertising Age* editorial offhandedly mentioned "the new online media" as a direct-marketing channel, at the same time noting that more typical "direct work" was hot "while the World Wide Web struggles with growth pains."[68] Soon, however, the meteoric rise of the internet and the rapid growth of technologies to help consumers escape advertising would pose challenges that traditional advertising could not meet. Marketers would look to direct response and product placement—newly rehabilitated and trendy for other reasons—for solutions. And as these activities moved the new imperatives of marketers forward, it would become clear that marketers and media practitioners were pressing for the most profound transformation of media and marketing's relation to American life in more than 100 years.

4 :: The Internet as a Test Bed

The story of marketers' online activities is a story of an institution trying to carve out a powerful place in a new medium. Marketers moved from a need to negotiate their very existence to a sense of hegemony over the new medium. Then came a period of heady pride that they could do anything they wanted in order to learn about internet users. It led to social opposition that forced them to pull back and reevaluate.

The trick lay in figuring out a way to lead consumers to reveal themselves to marketers online and give the marketers use of that information. Solutions that consumer groups, legislators, and members of the public found unacceptable threaded through the web from the mid 1990s onward. Out of those fights, though, mainstream marketers and media firms began to create an idea of what it would take to legitimately seduce the audience for information and attention. The internet, the most interactive of electronic media, has become a test bed for marketers' solutions. They built on the traditions of product placement and direct marketing and transformed both. Marketers typically try to use the enormous amount of data they have gathered about individual consumers to decide whether and how it is worth engaging them in relationships via customized email, ads, and other online presentations. Attempts to assess customer value almost always take place secretly for fear that consumers would be angry that companies hold such specific information about them. Yet when consumers look appealing, the companies surreptitiously mining their data work hard to convince them that their businesses are trustworthy for long-term association. In that sense, marketers' attempts to create trust and their undermining of that trust go hand in hand.

::

Before 1994, when the internet was primarily a text-based medium, its users were fiercely protective of what an *Advertising Age* writer in 1993 called a "culture . . . which is loath to advertising."[1] They sneered at the non-internet-based information service Prodigy, which at the time devoted as much as one-third of its screen to ads.[2] In contrast, the "usenet" discussion groups that drove that online world had created strong norms against the sending of obvious and persistent sales messages. People who went against the norms were "flamed"—subjected to barrages of angry replies. A *Boston Globe* reporter commented: "While television viewers are accustomed to being bombarded with advertising messages, on-line etiquette frowns on such displays."[3] Even in this environment, advertisers found ways to reach readers. The key, said those doing it, was subtlety. They often used the euphemism "information provider" and emphasized the helpfulness of the activity.[4] When the World Wide Web browser entered the scene in 1993, its ability to show both graphics and text made it an obvious place for advertising. Aware of the online anti-commercial tradition, an *Advertising Age* writer counseled caution. "Marketing on the Internet a Daunting Prospect," read one article's title. "Those who try must disguise their ads as services."[5]

There were some advertising executives who didn't think the internet would ever be a popular advertising vehicle. Advertisers, they held, wouldn't support it, just as they hadn't supported teletext ventures of the 1980s such as Viewtron, Gateway, and Venture One.[6] Among a new generation of interactive marketers who hotly disputed this view was Martin Nisenholtz, a senior vice president at the direct-marketing subsidiary of Ogilvy & Mather. He argued in *Advertising Age* that consumers had decided in favor of the internet and that marketers would suffer if they stayed away. "The evidence suggests," he wrote, "that the online community does not need advertisers to succeed."[7]

Rather quickly, mainstream marketers and their agencies came to accept that they would have to find ways to interact with consumers in the new media environment or else risk being shut out of major areas of social life. The idea was at the core of an instantly famous speech given in May 1994 by Edwin Artzt, chairman of Procter & Gamble, at a convention of the American Association of Advertising Agencies. Artzt said he still believed in the importance of broadcast television for reaching huge numbers of people at the same time to sell products such as "four hundred million boxes of Tide." Yet he felt it was important to consider a variety of methods beyond the major television networks to get the "broad reach" the firm needed. Procter & Gamble had already begun to use customer segmentation and target marketing. What worried Procter & Gamble's chairman primarily was not that new technologies would encourage more targeted advertising. Rather, it was the "chilling thought" that emerging technologies were giving people the opportunity to escape from advertising's grasp altogether.[8] Artzt noted that the personal computer could be "a formidable future vehicle for advertising and even programming." He said that CD ROMs and online services were "bound to produce major changes in marketing goods and services to the public." He reminded his audience that the advertising industry had worked in the past to get the media to meet its needs. He urged the American Association of Advertising Agencies and the Association of National Advertisers to move urgently to consider how new media would affect advertising and how they could be shaped to the advertising industry's benefit. "We may not get another opportunity like this in our lifetime," he said. "Let's grab all this new technology in our teeth once again and turn it into a bonanza for advertising."[9]

There were already people and companies trying to do that. In 1994, the website Hotwired was the first to create a pictorial advertisement—a "banner"—that visitors to a site would see immediately without further clicking. Helping marketers create websites and ads for the online population were small "interactive" agencies such as Modem Media and Onramp. They saw the internet as a place to create models of advertising for all future media. This approach often emphasized continuity of web advertising with traditional branding and image practices, albeit with the potential for interaction with customers.

Direct marketers, in contrast, saw the web through the lens of their business: as a new way to gather relevant names and then reach out to those names quickly and efficiently. They exulted that chat rooms and other postings allowed astute practitioners to troll for names of people who, by their comments, separated themselves according to different lifestyles and interests. One consultant exhorted marketers that "hundreds of thousands of names and addresses are floating on the internet, waiting to be listed, organized, sliced and diced." After all, he pointed out, "the internet is essentially one giant agglomeration of special interests." He suggested that an entrepreneur roam the internet searching for names, mailing addresses, or phone numbers that individuals displayed in chat rooms and computer bulletin boards devoted to particular subjects. The entrepreneur should, he said, transfer these names, addresses, and numbers to a relational database, where they could be linked to attributes inferred from the topics of the individuals' messages. "Suddenly, an [internet user] is silently captured in a database and will soon receive information through the mail tailored to specific interests. What was learned cruising the internet has been vacuumed and converted to a targeted selling proposition."[10] Such hegemonic aspirations led directly to spam—and to public anger. Spam (unsolicited bulk email) got that name from some internet denizens of the late 1980s, who seem to have named it in dubious homage to a Monty Python sketch in which characters prance around singing the name of the canned meat product.

Spam of the electronic kind began flooding users' mailboxes in geometrically increasing numbers in the mid 1990s, just after email came into widespread use. Increasingly sophisticated entrepreneurs often relayed their messages through computers around the world whose owners were not even aware that they were being exploited for that purpose. The same technology that made the internet a relatively inexpensive means of targeting individuals on the basis of particular demographic or lifestyle qualities also made it incredibly cheap to send millions of missives to virtually anyone. The spammers would realize profits if only a small proportion of the millions who received their messages mailed money or went to their websites to order the products they were hawking. The attacks included fictitious return email addresses that appeared legitimate, and internet service providers such as Juno and AOL found it nearly impossible to stop most of them. Firms using legitimate return addresses worried about the implications. "When it gets to 100 [spam emails a day]," Eric Arnum of the *E-Mail Messaging Report* asked in 1997, "will I even look in my mailbox?"[11]

If spam endangered marketing because it angered consumers over information delivered to their computers without consent, cookies put online marketing in jeopardy because of the information they allegedly could retrieve from consumers' computers. Although the idea of a cookie can be traced to 1992-93, the type used today on the web was developed in June 1994 by Netscape Communications.[12] A cookie is a collection of information that a website places on a person's computer—for example, the pages visited on the site, items the person put into an online shopping cart, and the person's user number, user name, and password—so the site can take that knowledge into account when the person returns. The controversial text file was a necessary first step in the eyes of many potential advertisers to get rid of an information equality built into the web: An online firm could not tell anything about a site visitor—even whether he or she was new or returning—unless that person wanted the firm to know. That bothered marketers. A 1996 *Advertising Age* article put it this way: "Ever since the Web gained prominence as a commercial medium, marketers and publishers have demanded some way to understand how users move through their sites."[13] This article pointed out that cookies "aren't able to grab an email address" or to probe an individual's computer. That may not have been understood by everyone who reacted with alarm to cookies' existence. Even many who did understand that cookies could not by themselves elicit a person's name or address acted angrily. One reason was the initial surreptitiousness of the activity. Although Netscape implemented cookies in late 1994, Netscape (and Microsoft, which enabled them to be used in its Internet Explorer browser) didn't make the existence of cookies or information about how to stop their insertion into one's computer generally known until about a year and half later.[14] Another reason for the consternation over cookies was a fear that they would lead to technologies that would help strangers secretly learn more and more about web users.

Circumstances justifying that concern came quickly. In March 1996, for example, *MacWeek* published an article about the ways in which a combination of JavaScript and HTTP Cookies on the Netscape Web browser for Apple Macintosh computers could be used to "retrieve a user's email address, real name and activity from the Netscape cache file, which documents a user's movements on the Web."[15] Netscape acknowledged the problem and said it was taking steps to remedy it and to make the cookie more secure. Nevertheless, such incidents and the very presence of cookies worried people that the new medium might threaten its users with theft of personal information. Representative Edward Markey (a Massachusetts Democrat), the ranking member of the House Commerce Subcommittee on Telecommunications and Finance, said bluntly: "The same libertarian quality that has stimulated such rapid growth of the internet gravely threatens to cripple its promise." The web, Markey added, "has spawned an exponential increase in commercial voyeurism that is tearing privacy rights asunder. . . . At risk is consumer confidence in the medium. When consumer confidence plummets, so will economic activity on the internet."[16]

Markey and other strong supporters of web commerce were worried these concerns would stifle what was to that point sluggish growth. Stores and manufacturers were pitching tents online; so were newspapers, magazines, television networks, movie studios, and various entertainment, news, and information operations not affiliated with other media. It was clear to these web publishers that American consumers, having learned in the past century that content would be cheap because it was supported by advertising, would hardly ever pay for content on the web. Consequently, they expected that their fare would be supported by advertising. In 1996, though, some telecommunication experts were already opining that the internet was being over-promoted as an electronic marketplace. One journalist observed that consumers were "refusing to pay for what they're getting for free."[17] Consumers' reluctance to go online for fear of losing control over personal information seemed like an additional problem that could kill what many still considered a huge potential commercial resource.

::

The initial phase of concern about consumer resistance passed quickly, though, and privacy issues faded a bit. At the end of 1997, a Forrester Research study estimated that online retail revenue would total a record $2.4 billion in that year, "driven in large part by new security technology, easier-to-use commerce sites and advertising that is helping to reduce consumers' fear about shopping online."[18] Observers were also noting that advertising spending was hitting new highs, now close to a $1 billion for the year.[19] To direct marketers, this portended a lucrative future. With the presence of cookies to track people's activities on the web and the ability of companies to reach out directly to customers or would-be customers through email and rich-media links, direct marketers increasingly saw the web as their turf. "In the great debate over whether the internet is a branding or direct-marketing medium, the DMers have clearly won," *Direct* claimed in 1997.[20]

Branding-oriented advertisers did not necessarily agree. They were encouraging an infrastructure to help them evaluate the internet as they did traditional media, talking, for example, about "targeted buys on women's channels and networks" as the percentage of women online soared.[21] Using new technologies such as Shockwave and Java, advertisers were going beyond the static banner to create ads that incorporated interactivity, electronic commerce, sound and animation. Central to these developments was the rise of third-party advertising networks such as DoubleClick, SoftBank, and Real Media. For a cut of the advertising fees, they used their own servers to send commercial messages for advertisers to thousands of websites.[22] Like direct marketers, they had an interest in tracking web users and gathering profiles on them so as to decide whom to target, when and with what ads.

The integration of brand messages into users' activities on the internet developed quickly. Advertising messages were piggybacked across free Juno and Hotmail email accounts. Zapme offered free web access on the condition that users viewed a flow of ads at the bottom of the screen. Firms as diverse as General Mills and Chrysler increasingly sponsored—and even bought product placement in—online games. "We want to be a ubiquitous presence, reaching consumers where they are," a Kellogg Company spokeswoman said in 2000 in regard to such activities.[23]

Also feeding into the desire for ubiquity, but below most people's radar screens, was the hiring of firms to surreptitiously insert brand names into chat and even instant message discussions to encourage buzz about products.[24] In 2005, Dei Worldwide, an important firm in this growing business, made the following claim: "We deliver your messages seamlessly, integrating them into the context of the conversations that are already occurring in these online communities."[25] Taking word-of-mouth in a slightly different direction, firms developed online clubs around lifestyles or products that aimed to encourage participants to talk up brands offline as well as on the web. Teenagers, who were flocking online, were an especially attractive target. By 2001, Procter & Gamble had started Tremor, a program that used the web to recruit youngsters between the ages of 13 and 18, collect personal information about them, then to solicit their opinions about certain products and services and their help in hyping the products they like.[26]

Many of these activities used cookies or other vehicles to track their users and collect data about them. But there was no denying that web practitioners with a direct background were thinking particularly systematically about how to extend database marketing approaches to the interactive environment. In the late 1990s email was taking off as a sophisticated direct-marketing tool, despite web users' increasing anger over spam. In 1997, Amazon.com gained credit for its continual reinforcement of its customers through email to them. Third-party email firms emerged to help advertisers with bulk emailing, much as bulk direct-mailing companies did in the postal world. Several pushed incentive programs in which people who signed up to receive advertising by email received points toward rewards.[27] Such activities could in 1997 be audited by companies such as I/Pro, a factor that further legitimized the sector with advertisers. Meanwhile, the number of customers signing up for such free advertising-supported email services as Juno surged.

A November 1999 article in the U.S. Direct Marketing Association's magazine was ebullient: "It seems like only yesterday that *Direct* was publishing stories about consumers—and businesses—not being quite ready to use email as a marketing tool. The reasons were many. There were no lists available. Consumers were wary of their email boxes becoming full of what would soon become known as spam. And marketers weren't quite sure how this new medium fit into their targeting arsenal." But "times have changed. Email marketing is here and rapidly becoming an important component in many companies' marketing mix."

Many direct marketers saw the rise of email marketing on the web as a validation of the move toward one-to-one marketing described toward the end of chapter 3. Perhaps the 1990s' hottest book on that topic, Don Peppers and Martha Rogers' *The One to One Future: Building Relationships One Customer at a Time*, came out in 1993, just at the edge of the web's arrival. The book didn't mention the internet, possibly because of that network's reputation as anti-commercial. *The One to One Future* did, however, use Prodigy as "the best current model" of the "interactive, dialogue-intensive marketing environment" that the authors advocated—even though it called the service "a surprisingly dull disappointment."[28]

With the huge increase of people using email on the web by the late 1990s, the goal of cultivating customers on a continual one-to-one basis seemed like an approachable reality. *Direct* noted that email services began to boom in 1999 when companies with websites realized they needed to drive traffic there and "figure out ways to capture and keep customers."[29]

But direct marketers' views of the internet would prove to be controversially connected to their insistence on bringing their offline information approaches to the new digital arena. The terms they used to discuss online marketing sounded solicitous about consumers' privacy, yet the meaning they drew from the words got advocacy groups and even government officials nervous. It led to a tug of war over the way people ought to be handled in the new media era.

The words "relationship" and "trust" were at the center of the imbroglio. They sounded friendly and reciprocal. In dubbing 1999 "the year of relationship marketing," *Direct* quoted the director of email services of one firm as saying that that to be effective, email marketing must build relationships and encourage repeat site traffic. It also quoted the head of IDG List Services as saying "When you're involved in creating a medium, you need to teach people how to respond. Building trust is important."[30]

But central to this discussion of relationships and trust was direct marketers' own awareness that the situation was far more complex than pursuing full integrity and reciprocity with desired customers. One firm's director of email services commented to *Direct*, for example, that long-term relationships with even desired customers had to suffer as a result of businesses' imperative for "sales, sales, sales."[31] He noted that at a time when at least one retailer was claiming a return on investment of more than 300 percent for its email efforts, many marketers found immediate gratifications from purchasers too good to pass up in favor of long-term commitments to them. In that context, the companies defined trust in ways that would help them sell quickly and burnish a bottom line. That tension was evident in the implementation of "trust" on the part of the head of IDG List Services who was quoted earlier. He was actually referring to web visitors' accepting the sale of their names to marketers through a negative option.[32] That is, the people did not check a box when they registered on the site or in email that gave them the opportunity to "opt out" of being sold. IDG pioneered this approach to getting names off the web as an alternative method to using the names of consumers without any claim to consent. But to internet practitioners who still thought of the web as a special place, opt out was a bad internet practice that ignored the medium's tradition of openness and preyed on people's tendency to ignore the boxes.

Web marketer Seth Godin promoted an anti-opt out philosophy in a book titled *Permission Marketing*. In it, he derided traditional direct and branding ads as "interruption marketing" not suitable for the digital world. Instead, he argued that marketers should get consumers' explicit authorization to reach out to them. The Canadian Direct Marketing Association agreed. It required its members to get consumers' consent before sending marketing email.[33]

The U.S. Direct Marketing Association strongly resisted the idea, however. Longtime American direct marketers viewed Godin's approach as naive. They liked opt-out because it had worked and continued to work in postal mail and phone marketing; they disliked opt-in because they believed it was not amenable to easy prospecting for new customers. "In the internet space," said one direct-marketing executive, "a lot of people are losing money with opt in. In direct marketing, a lot of people are making money with opt out." The head of the DMA agreed: "A pure opt in model is a little bit pragmatically and commercially unrealistic," he told *Direct*.[34]

::

The tough insistence by American direct marketers that the web should be treated like earlier analog media when it came to consumer's information came under increasing challenge during the late 1990s. Two controversies eventually forced them as well as advertisers using targeted banners to revise their rhetoric and some of their guidelines about consumer trust and customer relationships on the web. One of the incidents was a revelation that websites and web marketers were soliciting information from youngsters about themselves and their families. The other was the DoubleClick advertising network's purchase of Abacus, a database firm with information about millions of individual consumers.

Public contention over marketers' desire for children's personal information was ignited in 1996 by a report from the Center for Media Education titled "Web of Deceit." The CME found that many sites were using varied techniques, including surveys, contests and offers of gifts, to get children to provide such personal data as an email address, street address, purchasing behavior and preferences, plus information about other family members. Together with the Consumer Federation of America, it called on the Federal Trade Commission to develop guidelines for protecting children's privacy online.[35] Thus began a complex set of dances between direct marketers, the Federal Trade Commission, advocacy groups, and Congress around issues such as self-regulation, the age at which children should be protected, and the need for laws to dictate proper information practices when it came to youngsters and adults.[36] In mid 1997, the U.S. Direct Marketing Association, alarmed by bad public relations and the gathering momentum toward legislation, began a "Privacy Action Now" initiative to push marketers to quickly adopt their own standards for protecting consumer privacy. A year later, though, Chairman Robert Pitofsky of the Federal Trade Commission commented that his organization viewed "self-regulation as not working." He cited an FTC-sponsored study which had found that in March 1998 85 percent of 1,200 marketer web sites collected data, but only 14 percent provided data on their information practices. Of the children's sites surveyed, 89 percent were collecting information; while 54 percent were provide some disclosure about information practices, only 23 percent were asking children to seek parental permission before they give information. Furthermore, fewer than 10 percent of the sites were allowing parents control over any information collected from children. "On kids, we are not willing to wait [any longer] at all," Pitofsky said. "Our proposal is that Congress ought to act promptly."[37]

Pitofsky also warned that his patience was wearing thin when it came to adult websites' claims of adequate self-regulation. He was in part referring to the brouhaha over DoubleClick. The ad-serving company had for several years tracked individuals' activities on the web by placing cookies on their computers' hard disks. In 1999, when DoubleClick announced it had bought the database firm Abacus, many assumed that it had plans to link the information on its cookies to the names and addresses of millions of Americans.

Company officials initially stated that it was impossible to associate an anonymous cookie with personally identifiable information (PII) about someone in a database. Yet it soon became clear that DoubleClick had come up with a way to do it. The idea was to entice web users whose computers carried DoubleClick cookies to go the company's Netdeals.com sweepstakes. There, to enter the contests, the user would have to type in his name, his email address, and other information. That would create the needed tie between cookie and PII and allow DoubleClick to query abacus about the person's offline purchasing activities. Also involved would be the

Abacus Online Alliance, a group of sites that had agreed to share and to pass along to Abacus the personal data provided online by consumers.

To marketers who endorsed it, DoubleClick's proposed activity merely paralleled the kind of profiles direct marketers were gathering offline all the time without consumers' permission in order to create direct-mail lists to solicit business by mail and phone. They were puzzled when the proposed merger created a firestorm of controversy that energized the emerging community of web-privacy advocates. "DoubleClick is engaging in surreptitious collection of data," said Ari Schwartz of the Center for Democracy and Technology, "and consumers need to opt in, not opt out." Schwartz acknowledged that he and others were pursuing DoubleClick because it was the largest company that aimed to link online PII with an already existing database about people. He emphasized that other companies were already carrying out what DoubleClick proposed to do. One example was Naviant, a processor of consumers' online registration of computers, software, and peripheral equipment for more than 60 software and hardware vendors. Through the online registration, "they've got everything at that point," including name, address, email, and a cookied browser, which they use to target consumers with online ads, Schwartz said.[38] These disclosures moved privacy concerns into the general press. High-profile reactions to the revelations kept them in the news. Threatened with lawsuits, third-party advertising networks had promised to allow consumers to opt out of profiling and DoubleClick had backed away from linking Abacus to its internet data.[39] Still, there was concern in high places. Congress announced two privacy task forces. Senator Robert Torricelli, a Democrat from New Jersey, threatened legislation to force web sites to seek consumers' approval before sharing personal information.[40] Banner-oriented advertisers, worried that the new legislative agenda would affect their abilities to target their web ads, conceded through the American Association of Advertising Agencies that marketers needed to do a better job of "ensuring that consumers understand what is going on" and have a "better sense of control."[41] Marc Rotenberg, executive director of the Electronic Privacy Information Center, pointed to the larger stake for privacy advocates. "The critical issue is what will be the future ad model for the internet," he said.[42]

Caught between lobbyists and the indignant press coverage of advocates, the federal government went only part of the way to rein in marketers. In 1999, following the advice of the Federal Trade Commission, Congress passed the Children's Online Privacy Protection Act (COPA), which made it illegal to ask children under the age of 13 for personally identifiable information without first getting their parents' permission. The FTC wanted to make the cutoff age 16, but marketers pressured lawmakers to lower it. Congress also did not follow the FTC's recommendation to approve new privacy legislation along lines that Torricelli and others had proposed. Nor did the FTC and sympathetic lawmakers make headway to require websites to follow four standards that had been developing in privacy circles as basic norms of online behavior: *notice* of a site's privacy policy, the offer of *choice* to consumers about giving up information, *access* by consumers to their the site's information about them, and *security* for that information so that only the groups that consumers allowed would gain access to them.

The Direct Marketing Association and its constituencies continued to insist that self-regulation would result in a fair advertising environment on the web. Their collective reply to privacy advocates was that they were mounting activities to show consumers that marketers deserved their trust online. The DMA announced it would bar membership by marketers that didn't have a privacy policy in place by July 1999, and IBM said it wouldn't advertise on a site that didn't have one. Moreover, to gain public credibility for the policies, websites began to use inspection companies such as E-trust and BBB Online that offered seals to assure consumers that the privacy policies of sites they had audited were honest. The halfheartedness of these efforts was obvious to anyone who examined them. While announcing an intention to develop trust, many popular sites actually undermined it in order to continue collecting data from their visitors without making them nervous. The standard response to the FTC's desire for notice told the story. Websites still typically used the opt-out alternative, and they let visitors know as little as was possible about data-collection activities in as polite but complex a fashion as possible so they wouldn't really understand what was going on but would feel good about them.

A patterned privacy policy was central to this strategy. In the first paragraph or two readers learned that the company cared about their privacy and wanted them to read the policy. Several paragraphs then followed detailing how the site visitor would actually be giving up information to a variety of "affiliates" who were typically unnamed, and also, perhaps, to third-party advertisers who did not fall under the rules of that policy but instead had their own policies. The paragraphs that spelled out these terms were not only often written in turgid legalese, they were presented in no predictable order. A visitor would have a hard time knowing where to find a site's approach to a particular issue without trying to read the entire policy.[43]

Privacy policy acknowledgements that sites captured, shared, bought, and sold marketing data about individuals often were written so as to cover up those activities. For example, the standard statement that visitors would get access to the information they had given the site simply meant that they could check and revise data they had revealed to the site during registration. No mention was made of the conclusions the site would be drawing about them after tracking them and buying information about them from database firms. On the contrary, a consumer looking at the registration data might get the impression that a site kept only minimal knowledge about them. And the seals that sites carried often didn't mean what they seemed to say. A reasonable understanding upon seeing the E-Trust seal, for example, suggested that it meant the user's data would not be shared. A trip to the E-Trust privacy policy revealed nothing of the kind. E-Trust simply promised that the site did what the privacy policy said it did. A site could therefore sell all information it gathered and bought about visitors and still get the E-Trust seal.

::

Borderline misrepresentations and invitations to confusion may have kept some regulators at bay, but they did nothing to make Americans trust the state of their information online. Phone interviews of random samples of Americans coordinated by the Annenberg Public Policy Center consistently pointed to widespread concern about releasing personal data online.[44] In late 1998, for example, when consternation about the DoubleClick merger and about questions websites were asking youngsters was in the news, 77 percent of parents with online connections at home admitted they were strongly or somewhat concerned that their children were giving out personal information about themselves when visiting web sites or chat rooms.[45] A year later, the Policy Center research focused specifically on the privacy issue and found that 36 percent agreed or agreed strongly with the statement "I sometimes worry that members of my family give information they shouldn't about our family to websites." Moreover, 59 percent agreed or agreed strongly with the statement "My concern about outsiders learning sensitive information about me and my family has increased since we've gone online at home."[46]

In the context of Congress' setting the age for parents' consent for children to give personally identifiable information below 13, it is noteworthy that 61 percent of the adults agreed or agreed strong that "I worry more about what information a teenager would give away to a website than a younger child under 13 would." Perhaps less surprising was the almost uniform agreement (96 percent agreed or agreed strongly) with the statement "Teenagers should have to get their parents' consent before giving out information online." And then there was the set of findings from a comparison of parent interviews with responses from youngsters aged 8 to 17 that seemed to confirm adult concerns: A much higher percentage of children than parents said it is "OK" or "completely OK" for a teen to give out a variety of family information—from the vacations parents take to how much alcohol they drink—in return for a gift.[47]

While marketers' findings about consumers' privacy concerns paralleled those found in the Annenberg report as well as other studies, and while the Annenberg study received wide attention in the general press, marketers and internet firms interpreted the situation quite differently. Consistently, studies they sponsored ended up arguing that American society had become quite alert to the particulars of its information environment. Americans, they said, typically understood their information options and were willing to negotiate privacy demands with companies that could offer something in return.

Alan Westin's Privacy and American Business consultancy became an important promulgator of this notion that Americans were making rational cost-benefit analyses about whether to release their information online. Westin acknowledged that Americans were nervous that companies would not handle their information properly. He contended, though, that if people were assured that companies could be trusted to deal with the information honestly and keep it secure, they would consider sharing it. He argued from his survey data that 58 percent of Americans were "privacy pragmatists": "They examined the benefits to them or society of the data collection and use, wanted to know the privacy risks and how organizations proposed to control those, and then decided whether to trust the organization or seek legal oversight."[48] This description of Americans as wary but savvy and ready to trade their information for benefits became the major underpinning for the way many mainstream marketers began to approach their internet work in the early twenty-first century. It painted a picture of consumers' skills that justified supporting legislation requiring an opt-out rather than an opt-in approach to unsolicited bulk email. It also may have curiously justified the DMA's opposition to legislation that would allow private citizens to sue marketers for civil damages if they send them spam. After all, adept consumers on the prowl for corporate gold might well pursue not only the real bad guys but "legitimate" marketers who were "accidentally" emulating some of the spammers' tactics.

In its final form, the Federal Can Spam Act of 2003, which went into effect in 2004, did require clearly marking a message as commercial email, providing an electronic means of opting out, and publishing a snail mail address to contact each sender in order to opt out of receiving future emails—and it allowed state attorneys general or internet service providers to file civil suits. In a surprising defeat for opt-out advocates, the Can Spam Act also prohibited sending unsolicited commercial messages to wireless phones without the recipients' advance permission. The powerful cellular phone industry had objected to "opening the marketing floodgates on their users."[49]

Although Congress had aimed the Can Spam Act at fringe actors, *Marketing News* bluntly called it a "regulatory setback" for marketers.[50] Marketing attorneys complained about the act's ambiguity and proposed ways to try to get around its opt-out requirements. The Association of National Advertisers urged the FTC to oppose a do-not-email list that Congress proposed as a possible by-product of the law.[51] At the same time, the trade magazine quoted a DMA executive's belief that such legislation was inevitable and that the most likely alternative—a patchwork of incompatible state laws—was distasteful: "Fifty or more email marketing laws will cripple e-commerce."[52]

This was the balancing act that the marketing establishment seemed destined to carry out with every new technology that promised the ability to follow individual consumers across the digital terrain. Resistance by advocacy groups, worries among the public, pressure from state and federal lawmakers, and conflicting interests among marketers (such as cell phone firms) sometimes called for public compromises on definitions of access, notice, and choice regarding personal information in ways that could benefit consumers. At the same time, marketers saw the need to follow individual consumers across the digital terrain enlarging rather than diminishing. The reason was that consumers more and more had the power to desert them. "Companies must recognize that they increasingly have to engage gods and are not dealing with helpless consumers anymore," Rishad Tobaccowala, an executive vice-president of Starcom MediaVest Group, told an industry group. The specific reference was to technologies that he was sure growing numbers of Americans were using to whiz past or obliterate conventional TV and internet advertising. With a click of a mouse or a push of a remote-control button, the resources advertisers had invested to target consumers were wasted. "This," Tobaccowala added, "is particularly true of young people." That remark reflected media and marketing practitioners' worry that the upcoming generation of consumers that was particularly smart about how to avoid ads and yet maintain the benefit of advertising-supported media.

Hyperbolic though comments made in the trade press and at conferences often were, they ratcheted up pressure on marketing and media practitioners to carry out two sets of conflicting activities at the same time. They had to mine, store, and exploit consumers' information while at the same time encouraging them to provide information about themselves. Marketers had to violate the spirit of their compromises and assurances of information openness even as they hailed them and the consumer demands they represented. That contradictory imperative was central to a much-discussed February 2004 speech that Procter & Gamble's marketing chief, James Stengel, presented at a meeting of the American Association of Advertising Agencies ten years after Ed Artzt issued his call to arms in front of the same organization. Stengel emphasized that, although not all of Artz's predictions had come true (he had overestimated the amount the public would pay to get media fare without ads, for example), his concerns were essentially valid. "Consumers are less responsive to traditional media," Stengel noted. "They are embracing new technologies that empower them with more control over how and when they are marketed to."[53]

Stengel emphasized his belief that consumers would opt into the process. "All marketing should be permission marketing," he said, and "all marketing should be so appealing that consumers want us in their lives . . . and homes." To do that, he cautioned, required creative content and "connection points" in a variety of media and environments that "can have a profound impact on how we reach consumers beyond the 30-second spot (in-store, mobile-technology and text-message groups; pop-ups, digitized billboards that can be programmed; coffee wrappers)."[54] Accomplishing that, in turn, meant both knowing the consumer intimately and reaching that consumer in a way that would measure the effectiveness of marketing. In saying that Procter & Gamble would use permission marketing to do those things, Stengel tied his company to rhetoric that put it on the good side of web advocates and policy makers. The contradictory impulse lay in what he left out. Stengel did not discuss letting customers in on the information Procter & Gamble collected about them after receiving the permission and while it interacted with them. Few, if any, firms allowed consumers to know precisely how they tracked them, exactly what information they had about them, what they were concluding about it, and how they might use it to serve up ads and personalize content. Doing that would truly allow consumers to "opt in," to understand what stories were being told about them, why, and with what effect on the media they received. But in the decade-long wrangling over personal information online, including the DoubleClick and COPA battles, marketers worked so hard not to yield the most basic rights to access and notice that the deeper facets of data capture hardly came up.

To Stengel and other marketers, there were two ways to cope with consumer control in the new marketing environment: by encouraging consumers to interact with them, and by building a "permission-based" database about consumers to learn more about them and predict their purchasing behaviors. Although some might be tempted to read into these approaches the mid-1990s cheer that "the DMers have clearly won,"[55] the historical tension no longer made sense at a time when marketers had become fixated on Pareto's 80-20 rule as a rationale for searching out customers who would stay with them. In their 1996 book *The Loyalty Effect*, Frederick Reicheld and Thomas Teal had restated and updated Pareto's rule for marketers. *The Loyalty Effect* also pointed out that some firms actually lose money on customers until the second or third year of selling to them.

Many marketers resisted the idea that customers would genuinely have lifetime loyalty to a company or brand. More likely were cycles of loyalty that moved with changing phases of life. The most advanced marketers were looking for ways to combine the direct practitioner's traditional interest in databases and newfound desire for interactivity with the image advertiser's traditional emphasis on brands and newfound focus on product placement. The goals were to determine which individuals were likely customers, to learn what emotional and logical bonds would move them to buy the product at their particular points in life, and then to interact with them online and offline, through various media, in ways that would lead to trial and long-term use.

::

Media and marketing practitioners understand they are only at the beginning of a long journey toward the aforementioned goals. They see mass customization as a crucial element of that journey. When an advertiser or a media firm carries out mass customization, it distributes content to an individual on the basis of information the firm has about that person that it expects will make a difference in attention or lead to a purchase. The distributed content—perhaps an advertisement, an article, a program, or a song—can be created in real time when the individual contacts the company at its website or by email. Creation can take place by instantly grabbing bits of pre-coded blocks of material and arranging them according to rules for that person's profile (the particular characteristics or labels that the firm has attached to him or her). More often, mass-customized materials are fully crafted in advance and held for distribution to people depending on their profiles. For efficient customization, database companies may lump profiles together into groups, or niches, on the basis of statistical similarities. Even when that is done, some companies may divide their customer lists into hundreds of niches.

A basic kind of online customization takes place via one of the web's most popular advertising forms: search-engine marketing (SEM). Its guiding logic is that people sometimes are interested in buying products related to the words or phrases (for example, "digital camera") for which they search on Google, Yahoo, and similar sites. Sellers of particular products (for instance, digital cameras) bid money for a search engine to place their website links alongside the "natural" web search results for chosen words or phrases. The bids can be for certain times and certain periods of time.

Each time a web user clicks on the advertising link (transporting the click-er to the seller's site), the seller owes the website the agreed-upon money. A variant on this "pay-per-click" activity is contextual advertising. To carry that out, search-engine firms make agreements with websites that allow their software to read the pages of the sites and place ads at the side of their web pages when they find words that their advertising clients have chosen.

A less popular but growing approach to online ad customization is behavioral targeting. Companies that do it argue that it is a complement to contextual advertising, because a person's history of web activities can often suggest more about what kinds of ads a person might want than would the content of a page he or she is reading at the moment. One form of behavioral targeting, called adware, appears when companies such as Claria, When-U, and 180solutions insert software into computers as part of their owners downloading of "free" software, such as the music-sharing program Kazaa. The companies then track the web activities of their anonymous users, infer their interests from those activities, and serve ads to them on top of the websites they were visiting. For example, if Claria's Gator software noticed that a user was going to pregnancy sites and then to baby-naming sites, it would begin serving that user ads for baby prod-ucts. Hijacked websites and consumers with slowed computers were aghast that tactics to take over their domains were not illegal. Claria and WhenU claimed that their advertisers had a right to reach their audience in that manner and that the audience would find the ads relevant. So far, lawsuits have not settled the issue.[56]

The software certainly has gotten around. The McAfee virus-detection company found 11.4 million adware applications in March 2004, and 40 million over an eight-month period.[57] Major advertisers doing business with adware firms include Priceline.com, J. P. Morgan, Yahoo, Verizon, Merck, and T-Mobile, according to research by Benjamin Edelman, an economics doctorate student at Harvard University. "They advertise with them because it gets results," Edelman told *Investor's Business Daily*.[58]

The belief that ad customization based on a web user's online activities gets results has encouraged less controversial forms of behavioral targeting. Rather than send pop-ups to the people they are tracking wherever they are, some firms (including Revenue Science) use individual websites, and others (including Tacoda) use a network of websites, to send ads to segments of their audiences that exhibit behaviors across the sites (as noted by cookies) that are attractive to the advertisers. At this writing, these networks, like the adware firms, have chosen to know only the behaviors of the individuals, not their names or email addresses. At the same time, the firms have no qualms about planting cookies on an opt-out basis, because the industry and regulators have tacitly accepted that approach to anonymous behavioral tracking.[59]

Online advertising networks of websites do not merge background data about individuals with the information they store about their movements across the web primarily because member sites don't want to share knowledge about their users. For their own purposes, however, individual sites often show no reluctance to link behavioral tracking of their visitors to demographic and lifestyle information they have requested at registration or bought. They use the combined data primarily to customize the sending of ads to individuals on their sites. Linking specific behaviors with particular ads can often take place instantly. Tacoda, for example, notes that its clients can "serve ads during a user session capitalizing on behavior that strongly suggests readiness to buy or take specific action." Tacoda also says that its tracking techniques allow advertisers on a site "to instantly categorize and target visitors by their brand preferences."[60]

Another approach is to use a marketer's in-depth knowledge of an individual's behavior and background to send that person customized messages. Procter & Gamble does that with its 240,000 teen Tremor members. Until a few years ago, Procter & Gamble did not allow voluntary membership in Tremor; Tremor's website noted that P&G searched for and picked its members. Perhaps in response to criticism of the activity's exclusivity, Procter & Gamble opened membership via the Tremor website to every person between the ages of 13 and 18 who wants to "join a group that likes to be heard." Nevertheless, the teen is asked to fill out an online questionnaire that tries to assess his or her influence on friends. The small print in the FAQ area of the site states that some teens will be asked to participate more often than will others, though it is vague on the reasons. The site's privacy policy also notes that Tremor "may supplement the information we collect with data obtained from third parties." Despite the welcoming image, then, a careful reading of the site's small print suggests that Tremor's database is used to select teens in a secretive manner while not appearing to do so.[61]

Database-driven selection is also beginning to emerge as a means of serving different brand placements in internet video games. In late 2004, inGamePartners, a technology firm founded that year, touted its ability to seamlessly integrate products into any gaming platform with an internet connection—a mobile phone, a BlackBerry, or an internet-enabled console such as an X-Box. Moreover, it claimed to be able to vary the ads a particular user sees on the basis of the user's geographic location (as derived from the connection the user's computer has to the internet). That ability allowed it to sign ad-distribution deals with GriffinRun and PHXX, two of the largest public online video gaming networks on the web. It is not difficult to see how "permission-based" use of personal information collected at registration could be used to target individuals further by placing different brands in different individuals' games depending on what kinds of people advertisers want and when they want them. This capability, *Media Daily News* noted, "will be breaking new ground in video game advertising."[62]

Another approach to such selectivity—one that fits the permission philosophy quite directly—is to encourage a favored customer the opportunity to make the decision to get messages. Dotomi Direct Messaging gives consumers the ability to opt in to receiving messages from a merchant in the form of banner ads on websites they visit. Say, for example, a woman named Rachel agrees to have The Gap send her coupons or notices about special sales in ads that may appear on one or more of the many websites she reads. The next day, as Rachel browses through WashingtonPost.com, a banner ad from The Gap appears that is directed to her. It offers her $10 off the price of a cashmere sweater to go with the pants she bought the last time she was at the online store. Later in the day, The Gap reaches Rachel at another website to tell her of a forthcoming online sale of the jeans she typically buys.

The ads are triggered through Dotomi by a cookie in Rachel's computer that is activated by Dotomi on the site. The customization of the message, though, comes from an analysis of Rachel's data at The Gap database, which is available to no one but The Gap. Clicking on the banner takes Rachel to the Dotomi Direct Messaging Center to retrieve other Gap messages, change her interaction preferences with the store, forward the message to a friend, or opt out of the process. Dotomi shares revenues from its clients with the websites on which the ads appear.

When executives describe such scenarios, they assert a fulfillment of fundamental marketer and media needs at the threshold of the digital era. Dotomi's trademarked tagline "Consumers Rule!" plays on the fear companies have that consumers will tune out their general ads. The phrase also feeds marketers' growing obsession with taking control back from the consumer (while asserting consumer power) by beginning with a predictive understanding of customer activities. John Fedderman, the president of Dotomi, places his firm's work squarely in the camp of customization technologies that "dig into your existing customer data" to get consumers to pay attention to ads again, develop special trust with groups of advertisers who promise not to share their information, and create "personal message" channels through which "marketers and consumers can build and retain life-long relationships."[63]

Major email service providers recite the same permission-based relationship mantra. A 2004 Forrester Research report on a national survey noted that, despite a flood of unwanted electronic mail "souring [consumer] attitudes toward email marketing," nearly 80 percent of consumers subscribe to messages from companies, and that they receive an average of 30 messages per week "from these permissioned senders." Forrester also examined data from four leading email service providers (DoubleClick, Bigfoot Interactive, Digital Impact, and Yesmail) and found that consumers opened and clicked on a link to a website for more information at the same rate in 2004 as in 2003. In fact, Forrester found that the four large email firms saw 37 percent of their permission-based messages opened, with a 5.1 percent click-through rate.

These sorts of generalizations are music to the ears of email senders. In tune with the buzzwords of digital marketing, the cutting-edge approach is to use email not as a tool for offers that marketers supposedly want to push at people but as a vehicle for sending customized offers that customer are likely to "pull" toward them—that is, open eagerly—because they reflect sophisticated analysis of their backgrounds and previous purchases. And, in fact, the database industry has responded with an armamentarium of methods that claim to accomplish this goal for online marketing. Epiphany, for example, advertises that it "offers a complete solution for optimizing interactions with customers over online channels such as the web, email, and SMS." The core idea is to bring together all the data a client firm has collected (and continues to collect) about its customers, analyze the data, create profiles of the individuals, and, by such methods as scoring them on various characteristics, customize interactions with them in profitable ways.

In a "case study" presented on its website, Epiphany claims that by using its expertise and software American Airlines has gained "a comprehensive view of its customers across all [electronic communication] touchpoints . . . to enhance customer relationships."[64] Examples:

• Every Tuesday, the airline distributes an email message to customers who have opted to receive special web fares and offers to select cities in the United States. The idea is to sell excess flight capacity as flight time draws near as well as to make customers feel good about the airline.

• On American Airlines' website, AA.com, Epiphany implements personalization and content management software to analyze customer profiles as they move through the site and then proceeds to "match them to relevant content and offers on the site."[65]

• AAirmail is an electronic newsletter sent to millions of customers that provides customized content and offers tailored to the individual profiles Epiphany has created. For example, newsletter articles vary to help individual customers reach their next top-tier status—Gold, Platinum, or Executive Platinum.[66]

As Epiphany and American Airlines see them, these activities are parts of what the airline's head of customer-relationship management calls "a unified view of customer behavior" that allows the company to "integrate data about past transactions and interactions, online or otherwise."[67] In fact, as this quotation suggests, increasing numbers of companies are going beyond the digital realm and using Epiphany or larger database firms such as Oracle-PeopleSoft or Acxiom to create central customer databanks for the instantaneous use of all customer information. As one writer put it, the repositories "collect data from all points" and then "tailor permission-based offerings to accommodate customers' finely segmented demands, wherever they originate."[68] A researcher in IBM's Industry Solutions Group emphasizes the need to look at "the patterns of a customer's activity, such as the types of products she likes, how she responds to promotions, and her price sensitivity" both online and offline. When a company has that kind of information about its customers, says IBM researcher Edwin Pednault, it can begin to ask "How are my actions motivating them to change from one [buying] state to another?"[69] In a similar vein, Acxiom tells its clients: "The ability to best serve your customers when it matters most—during the interaction—is critical to achieving customer growth and retention goals. Acxiom's customer recognition solutions enable companies to distinguish customers accurately and consistently, providing complete and instant access to relevant customer data across all channels of communication."[70]

The internet has become a crucial node in such contacts. Intrawest, a global resort firm, embraced this approach in 2002, moving over the next two years from annually implementing 50 rather undifferentiated direct-mail campaigns to mounting 2,400 customized email and direct-mail efforts during a twelve-month period. "Our customer now allows us to look at the customer holistically," Intrawest's head of customer-relationship management told *Direct*. "We can pull everything together and get a better understanding of what kinds of vacations they want." Now, she noted, the company would be more likely than previously to learn particular preferences of its customers—for example, that an avid golfer who live in Phoenix and travels to an Intrawest golf resort in Canada might be interested in some of the firm's resorts closer to home.[71] The individual profiles that Intrawest creates even include the channels—email, postal mail, telephone—that individual customers want the firm to use when communicating with them. Such multi-channel marketing is another tactic of customer relations that is accelerating with the advent of holistic databases.

The most important goal driving all these activities—for Intrawest and most marketers—is to discriminate. It is to digitally find, profile, and customize for the 20 percent or so "best," "high-value," or "prized" customers. The consultancy Acxiom puts the point bluntly for potential clients: "You need to know how to turn customer relationships into incremental return on investment; however, to succeed you have to have best-of-class customer and information management solutions."[72] Acxiom adds that its "prospect marketing solutions help you target consumers who share the same characteristics as your best customers." That includes "making marketing decisions based on up-to-date credit bureau information" and managing "multiple, concurrent marketing campaigns" that "personalize offers to prospects most likely to become high-value, long-term customers."[73]

The word has certainly been getting out. Forrester Research urges its readers to "focus on high-value customers" in customizing channels to win loyalty.[74] Procter & Gamble boasts that it has used database marketing and then email programs, websites, and online ads as well as traditional analog media to cross-sell to the 10–20 percent of consumers (what P&G calls "golden households") who account for most of the sales of its nine health-care brands, including Crest, Metamucil, and Nyquil.[75] Similarly, Verizon Wireless tells of routing high-value customers to the most senior agent available. An agent working with a high-value customer may be automatically prompted by a different script, allowing them to spend more time with that customer and to converse in a friendlier way. Gifts and discounts as well as attempts to "upsell" often accompany these interactions.[76]

The logical corollary to this favoritism, of course, is that less attractive customers should get lower levels of service than attractive customers get, or even no service at all. A 2004 Forrester Research report carries this idea forward, noting that financial companies, which arguably have the most sensitive and actionable data about people as customers, "have started talking openly about 'firing' unprofitable customers."[77] Forrester suggests that while getting rid of bad customers is certainly an option for all firms, two "more strategic" choices are "understanding what makes [such individuals] unprofitable and identifying ways to move them to a more profitable phase" and "servicing low-value customers as inexpensively as possible, generally through routing rules and channel options."[78]

::

The internet became the primary development site for mass customization in support of activities aimed at treating customers differently (or even pushing them away) depending on assessments of their desirability. Though privacy issues occasionally rose to challenge marketers' new ways of working with consumers, marketers learned to work around them, finessing government agencies with limited disclosures and protective rhetoric. Customers might be wary, but they were giving up their data to be in the game and get promised benefits. In 2004, *Direct* pointed out that the application of direct-marketing concerns to the digital world has created "a sea change for marketers, both in mindset and in the way they collect, store, view and use customer information."[79] It has, *Direct* noted, changed the way consumer-oriented companies selling everything from package goods to electronics to cars reach out to customers.

Google, Yahoo, CNN.com, and many other web firms encourage users themselves to "personalize" the look of the sites and the types of stories they get. One reason the sites like personalization is that it provides them with information that triggers customization. When users' indicate preferences for certain content, it may lead sites to cast up "contextual" ads that reflect those interests or the lifestyles they imply. Not only do many websites track and use those preferences; they combine them with an enormous amount of other data that the firms get about their visitors by tracking their movements through site pages and sometimes even by purchasing information about them from data brokers. Google doesn't appear to buy third-party information, but it certainly gathers and stores an enormous amount of information about what individual visitors do. Moreover, the company is increasingly linking all services for which users register— Froogle, Orkut, Google search history—to the user's gmail login account. As the journalist David Vise notes, "few Google users realize . . . that every search ends up as a part of Google's huge database, where the company collects data on you, based on the searches you conduct and the websites you visit through Google. The company maintains that it does this to serve you better, and deliver ads and search results more closely targeted to your interests. But the fact remains: Google knows a lot more about you than you know about Google."[80]

Even more intriguing as a basis for customized ad-serving are the growing numbers of sites that base their content on the individualized creations of their users. My Yahoo, My Google, and more specific web arenas such as the photo-sharing site Flickr (owned by Yahoo), the music-oriented "social-media" site MySpace.com (owned by the News Corporation), and the college and high school social site FaceBook all encourage users to personalize their material so that they can reach out to others like themselves. Read the privacy policy (which, research shows, few people do) and you will see that the personalization also provides grist for company databases that can be used to customize advertising and other forms of marketing communication.

Facebook's privacy policy, for example, says: "When you register on the Web Site, you provide us with certain personal information, such as your name, your email address, your telephone number, your address, your gender, schools attended and any other personal or preference information that you provide to us." The document adds that Facebook "also collects information about you from other sources, such as newspapers and instant messaging services. This information is gathered regardless of your use of the Web Site." Precisely how the company uses this stored information is impossible to glean from the privacy policy. Facebook cryptically notes that it "may provide information to service providers to help us bring you the services we offer. Specifically, we may use third parties to facilitate our business, such as to send email solicitations. In connection with these offerings and business operations, our service providers may have access to your personal information for use in connection with these business activities."[81]

What is taking place through this process is the transformation of individuals' personal creations or relationships into grist for customized material sent to them by site owners and their affiliates. Until recently, it has been advertising rather than news, entertainment, or information that has been custom-created by companies in response to individuals' database information. That has been the case with "consumer-generated media" such as Facebook as well as with more traditional "publishing" sites such as BusinessWeek.com. BusinessWeek uses Tacoda software to customize its ads based on user information, but it has not served different articles to readers to customize them for readers on the basis of their database profiles. Yet, according to Tacoda executive Kurt Viebrans, that sort of discrimination among readers is about to happen. In the early 2000s, Viebrans pointed out, web publishers focused on customizing their advertising spaces because that was where the most immediate revenues lay. Viebrans noted, however, that in 2005 the *Dallas Morning News* and other publishers began to talk quite seriously with his firm about applying its methods to editorial matter. Viebrans also acknowledged that, although Tacoda had enough work to do in the internet industry, many of the new direct-marketing strategies that had been developed online were beginning to migrate to digital television.[82] Would television accommodate them?

5 :: Rethinking Television

In his 2004 book *Madison & Vine*, Scott Donaton of *Advertising Age* criticized obvious product placements: "Those integration efforts that are forced will stand out like sore thumbs and will be rejected." The best product placements, he said, are "subtle and seamless, and appear natural to the audience. Those that work will begin with the consumer in mind and with the goal of creating compelling content, but will still manage to meet the needs of both the advertiser and the creators of the content."[1]

Donaton's suggestion was part of his "six quick rules that will serve as guidelines for the development of the Madison & Vine space"—that is, for the integration of brands into Hollywood products (television, movies, DVDs, and video games). Although in the book he went out of his way to deny it, Donaton's advocacy of "subtle and seamless" product placement contradicted a position he had taken in *Advertising Age* in 1999. His object of scorn then was a decision by the CBS television network to sell replicas of jewelry worn by characters on the soap opera *Guiding Light*. Contrary to his later retelling, Donaton did not merely condemn the idea because it was a badly conceived "gimmick." He generalized that "the dangers of blurring the line and destroying what remains of consumers' trust in media are growing daily. It happens in magazine publishing and with alarming regularity on the Web. Now it's network TV. The fundamental advertising model is at risk."[2]

Randall Rothenberg, an *Advertising Age* columnist, understood in 1999 that Donaton was criticizing product placement quite broadly. He needled his boss for not fully grasping the need for change in the television industry and product placement's role in it. For Rothenberg, the right way to see CBS's awkward placement-and-selling act (and a similar one by NBC) was "not as variations on infomercials, but as early yet necessary steps in the direction of electronic commerce." He went on to suggest that the networks' survival in an era of splintering audiences and rising costs would require a new symbiosis:

Just as marketers need to learn a thing or two about service and entertainment to draw audiences, so too must programmers seek new ways of aiding marketers (and themselves) in the sales process.

In the era now ending, they've been limited to using entertainment to attract viewers, who are then sold as a lump commodity to advertisers. Lucrative, but inefficient. The new model allows them to join (in differing gradations) information, entertainment, service, advertising, data mining and direct sales.

While the technology is still cumbersome—involving, as it does, telephones and operators—it's only a step away from the point and click ease that broadband will provide.[3]

Five years later, with TiVo, pop-up killers, and other technologies allowing consumers to circumvent the traditional commercial-as-interruption model, Donaton had clearly come to accept product placement as integral to the future of television and other media. He now agreed with Rothenberg that the rebirth of the old activity marked the start of "a fundamental transformation from an intrusion-based marketing economy to an invitation-based model." He didn't directly address how the "seamless" and "natural" product integrations he advocated could also be invitations if they were so seamless that people did not recognize them as appeals. He simply suggested that product placement should be considered a part of a system of "changes in how marketing communications are defined created, distributed and consumed."[4] That gets back to Rothenberg's insight that the really new aspect of product placement in the twenty-first century would be its linkage to the mindset of direct marketing. As I noted in chapter 4, that mindset, as it developed in relation to the internet during the 1990s and the 2000s, shifted away from requiring direct sales. It came to mean, rather, a database-marketing approach to consumers that involves *screening for appropriateness, interactivity, targeted tracking, data mining, mass customization*, and the cultivation of *relationships with individuals* based on those activities. Movements that reflect this logic are emerging, and some of the pieces are beginning to come together. It is clear that influential executives are beginning to accept bringing the mindset of direct marketing to TV as much as they have accepted it online.

In the 2000s, the audience's new abilities to interact with the TV set so as to get rid of commercials have been a frequent topic of discussion among television marketers. At a 2005 trade conference, the ABC television network's head of sales called for TiVo and other DVR firms to eliminate the rapid-forward button, so as to force viewers to use old-fashioned methods of ignoring commercials.[5] A bit more forward looking were the comments of a media buyer at the same trade conference. She exhorted broadcast and cable television executives to make shows' commercial breaks—the "pods"—unpredictable in length and time. That way, viewers wouldn't be so confident about how long they should leave the room for a when a commercial appeared, and they would end up watching more commercials.

TiVo tried a different compromise between viewer autonomy and industry resistance when it pitched to advertisers the ability to place a fast-forward icon or billboard on the screen to get the attention of the high percentage of TiVo users who sped through commercials. "It will be an opportunity for the advertiser to create a speed bump to get another chance to bring the person back into the commercial," said the firm's director of advertising.[6]

Many marketers viewed traditional product placement as the most attractive way of projecting a brand's personality. Product placement in a narrative that is compatible with the brand and in line with the desired audience's viewing habits seemed an obvious solution to the problem of ad-skipping. In fact, some executives began to promote product placement as part of a cross-media strategy that could be particularly potent if the brand were truly integrated into the entertainment to the extent that it was a crucial part of part of the action. That, too, wasn't a new idea; the Spanish-language network Telemundo had offered that possibility to ad agencies in 1989.[7]

The English-language television networks disdained such activity through the early 1990s, but eventually their fear of digital video recorders led them to reverse their position. Though DVRs were of only minor importance in the early 2000s, ad people believed projections that their presence in American homes would increase dramatically with the spread of digital cable. In mid 2004, Forrester Research predicted that between late 2004 and 2009 the percentage of American households with DVRs would rise from about 6 to 50.[8] More conservatively, Kagan Research estimated in late 2004 that the percentage of American households with such devices would rise to 32 by 2009 and to 49 by 2014.[9] Either way, TV networks began to promote product placement as a way to prepare for a future in which ad-zapping would be much more prevalent. NBC went so far as to say that in some of its series products would be so much a part of the narrative action that they could even appear in promos—a process the network proudly called "A-to-Z integration."[10] One episode of the NBC "reality show" *The Apprentice* involved a competition to design a bottle for a new soft drink called Pepsi Edge. Procter & Gamble's Crest toothpaste also figured in *The Apprentice*, as did Coca-Cola in the Fox series *American Idol*. Products also figured in fictional series, such as the Sci Fi cable network's *Five Days to Midnight* (Nissan), NBC's *American Dreams* (Campbell's Soup), CBS's *CSI* (MapQuest), and The WB's *What I Like About You* (Clairol Herbal Essences).[11] Product placement became so popular that major Hollywood talent agencies got into the business, Nielsen and other firms began to audit placements, and the Intermedia Advertising Group began to conduct regular surveys to establish what placements were recalled the most.

Coca-Cola marketing chief Steve Heyer became a standard bearer for the idea that product placement in TV entertainment ought to be understood as part of a "new way to reach and motivate our consumer" across media. "It's movies, music, video games that become a component part of our communications strategy and plans," he said at *Advertising Age*'s 2003 Madison & Vine Conference. Heyer emphasized the importance of knowing the "cultural references" that would move customers so that marketers could "manage the quality of our consumer relationships." Looking at people that way, he added, leads to the realization that "each person becomes a commercial market. And any agency that thinks a jingle connects like real music or a powerful movie and doesn't collaborate [with music and film producers] is lost."[12] The British trade journal *Marketing* noted that Heyer's speech had put "on the agenda" the need to make entertainment content central to a firm's communication strategy to "engage the restless, media-bloated . . . consumer."[13]

However, some marketing executives saw the kind of product placement Heyer was advocating as an unrealistic solution to ad-skipping. They pointed out that revenues from product integration could never replace the revenues that networks and producers make from TV's 15- and 30-second spots. They noted that companies often need more time to explain products and to create or deepen their brands' personalities. They also speculated that obvious product placement might "turn off" the new generation of consumers. "The younger the consumer, the savvier they are," a Comcast cable advertising executive asserted. "You just get hazed if you get heavy-handed when you try to appeal to that group."[14]

This is where linking product presentation with audience interactivity comes into the picture. Marketers are increasingly aware of viewers' active ability to "reach out to" product information while viewing. The near-term plan is to enhance interactive TV technologies that allow viewers to "reach out to" sales messages that interest them. Further down the road are applying databases to the interactive technologies and customizing the commercials that different viewers see. Up for debate is whether these approaches should be reserved for the separate advertising space surrounding TV's entertainment narratives or whether interactivity and database targeting should also apply to product placement in the shows.

The traditional approach to interactivity has been to keep ads and narrative entertainment separate. In fact, much of the early interactive television seems to have been driven not by a desire to sell products but by a desire to involve viewers with the programming in the quest for higher ratings. The hope was that the audience for advertising would get a boost along the way. Broadcast historians consider the 1950s Saturday morning show *Winky Dink and You* the first attempt at an interactive TV narrative. By mail or at stores, parents were supposed to buy clear plastic sheets for their children to stick onto the TV screen. When the show's cartoon character was about to fall off a cliff, for example, the viewer would stick the sheet onto the screen, use a crayon to draw a bridge, and watch the character cross it. (People who remember the show tell stories of parents who got furious at their kids for not using the plastic screen and drawing directly on the TV.[15])

Though the *Winky Dink* approach remained a unique version of viewer response, the basic idea of audience interaction with TV narratives continued sporadically in two directions, one involving pay-per-call 900 numbers or web-response addresses and the other involving the vertical blanking interval (VBI). Phone-TV interaction was relatively primitive. The three broadcast networks used it as a way to stimulate viewers' interest by encouraging them to phone in and vote for plot endings. NBC was particularly active in this. In 1982, *Saturday Night Live* became one of its first shows to poll viewers. Other applications on the NBC network included asking viewers to vote for which ending they wanted for an episode of *The A-Team* and which episodes of *Miami Vice* they wanted to see rerun.[16] Promotion of this sort of audience choice continued into the 2000s, though its implementation changed with the rise of the web. In 2001, NBC encouraged viewers to go to its website to vote on the ending to an episode of the comedy *Just Shoot Me* up to halfway through the show. (Different endings had already been shot.[17]) Three years later, the shows in NBC's *Law & Order* franchise—*Law & Order, Criminal Intent,* and *SVU*—asked viewers to vote for endings in different ways. In the case of *Law & Order,* audiences on the East Coast witnessed a criminal's escape from police detectives, while West Coast viewers saw her dead. Visitors to the network's web site could then see both endings and vote whether the character would live or die. On the series' next airing, the storytellers revealed that she didn't die. According to the web site, there were 62,074 votes for the character to live and 54,224 for her to die.[18]

Producers and network executives saw such interaction primarily as a way to whip up interest in the plots and to reward fans.[19] The same can be said for voting on beauty and music contests, such as NBC's 1980s program *The Most Beautiful Woman in the World* (which took votes by means of 900 numbers) and Fox television's *American Idol* (which, in a famous product placement, mentioned that that its public polls were conducted by means of AT&T's and Cingular's cell-phone text-messaging systems). CBS raised the stakes on this sort of judging when it applied the tactic to its "reality show" *Survivor* at the end of the 2004 TV season. The network encouraged viewers to vote for their favorite *Survivor* character online or by cell-phone text messaging. The winner was to get $1 million. People could vote as many times as they wanted until about a day before the next episode, which the network dubbed "America's Tribal Council."[20] The network announced that viewers had cast more than 9 million votes and that the winner had received 85 percent of them.[21] Despite the interactivity, none of these high-profile activities provided ways for viewers to directly reach out to sponsors in order to learn more about them. Two different ways to get viewers to interact with products in or around programming have, however, developed over the years. The more popular one involves "long-form" advertising—entire shopping channels on cable and broadcast TV as well as "infomercials" for which some cable and broadcast channels lease time in periods of low viewing. These programs display and describe the items to be sold and invite viewers to buy them by phone. Less widespread, but more important for TV interactivity in the long run, have been inventions that invite viewers to reach out to sponsors directly via their television sets.

In the early 1970s, engineers began sending data through the vertical blanking interval (VBI) of the analog signal. The VBI is the black stripe at the top and bottom of the TV picture. Broadcasters can use part of it to send data that viewers can receive using a special decoder. In the United States, the most common application was for text captions, often used by deaf viewers. But two VBI technologies made inroads into more than a million cable and DBS[22] homes in the late 1990s: Wink and WebTV. Both used the vertical blanking interval to send what Wink called "enhanced broadcasting." At the height of their involvement in VBI programming, in 2000 and 2001, Wink and WebTV provided hot platforms for companies creating content. Viewers with WebTV could play along with game shows such as *Jeopardy* and could learn more about Learning Channel shows such as *Trauma*. ESPN and CNN used Wink with similar capabilities. When the icon "I" appeared on those and other channels, viewers could click on it to request more information from a program's producer.

Wink and WebTV fell from favor as satellite and cable firms turned to using DVRs instead of VBIs to store materials and distribute them to viewers. The idea was essentially the same: to stream a variety of program possibilities from the satellite and then to allow the consumer to choose from among them. When a consumer wanted to respond directly, the response would be carried by the consumer's phone line. In the United Kingdom, the News Corporation's BSkyB satellite operation pushed the interactivity of the set-top box to the point that the "red button" on the remote became part of the national consciousness. It could provide up to eight simultaneous windows on the television screen, allowing people to watch the news with sound while looking at a weather forecast and viewing a football game.[23] A particularly popular use was for sports and live events; the viewer could select from multiple camera views that streamed into the set-top box. Targeted advertising and specially designed content also were practicable. For example, a viewer could select from among four commercials or two endings to a television movie. The software in the set-top box would remember the selections and could replicate the same type of request in the future without being reminded.

To what extent American viewers ultimately care about interactivity of the sorts described above became a subject of lively discussion among executives. In the late 1990s, Forrester Research and other consultancies were touting interactive television. Joshua Bernoff of Forrester Research was one of the most enthusiastic boosters of the embedding of commercial material in narrative programming. His report "Smarter Television," issued in 2000, got "top of the week" attention in *Broadcasting & Cable*.[24] Bernoff confidently predicted that by 2003 cable and satellite operators would roll out sophisticated DVR-laden set-top boxes. Television networks, he said, would then embed in their programs information that would allow viewers to learn about and buy the products they saw on the screen. For example, a viewer would be able to learn more about the sweater worn by Rachel, a character in *Friends*, and even buy it through the television, while still tuned in to the show.[25] Not only would "commerce on the screen" through such interactive video tactics amount to more than $11 billion by 2005; it would change TV's content, especially on cable networks aimed at niche audiences. "As traditional advertising becomes less effective," Bernoff said, "cable networks will swing in the direction of content well-suited to interactive response."[26] Bernoff suggested that the major networks' sitcoms and dramas might "drift in the direction of the MIT Media Lab's 'Hypersoap' demo, in which viewers can buy every item the actors are wearing or using."[27] Database-driven targeting would inevitably become a priority. "The masses of data collected through smarter TV will demand high-powered analysis for targeting. . . . As targeting becomes more effective, new ad-buying behaviors will arise—like American Airlines targeting customers who click on United ads." And for the less well heeled, "cable operators seeking an egalitarian image in low-income neighborhoods will offer . . . dollars off the cable bill in exchange for commitments to click on commercials. Click rates may go up around the end of the month as household members recognize they must click now or end up paying for laziness later."[28]

What Bernoff predicted certainly didn't happen by 2005. In fact, in 2005 no one—not even Bernoff—was predicting it would happen soon, even though interactive television took a huge leap toward general availability that year. In 2004 the News Corporation had purchased the American DBS company DirectTV. Imitating the News Corporation's DBS strategy throughout the world (most notably with BSkyB in the United

Kingdom and Foxtel in Australia), DirectTV announced that it planned to offer interactive services with its NFL Sunday Ticket package. That included the ability to check scores, statistics, and fantasy-team developments with a click of the remote. A few months later, the number-one and number-three American cable system operators, Comcast and Cox, formed Double C Technologies, which then purchased the North American assets of Liberate Technologies, a nearly bankrupt California-based developer of interactive television software. Though they denied that the collaboration was a response to the News Corporation's purchase of DirecTV, their announced aim was to allow customers to do what BSkyB customers could do interactively and more.

Comcast's executive vice president of new business development gloated that the interactions offered by cable companies would be far superior to those offered by DBS operators. Cable operators had the benefit of being able to send specific content directly through a two-way broadband wire into every home, while the satellite firms could only download content a person could choose from the set-top box. He pointed out that Comcast customers were already interacting with their cable company when they pushed their remote button to rent movies. Liberate Technologies' software, he said, would allow far more complex interactions. Examples he offered included purchasing products along with seeing who is calling on the telephone, checking email and voice mail, looking up sports scores and statistics, voting on a town-meeting initiative, and viewing the status of one's stock portfolio. In fact, however, the Comcast executive's notions of interactive television were hardly much beyond Wink and BSkyB and far less adventurous than what many of his customers were probably already doing on the web.

Marketers wondered whether viewers wanted that kind of interactivity while watching television. Stating that interactive television commerce had not been terribly successful on BSkyB, a U.K. analyst opined in late 2002 that "for many, the TV remains a 'lean-back' medium, though which people want to be entertained." The vice president of business development at Visible World, a new-technology advertising company, said in 2005 that Americans felt as the British did about interactive entertainment. Research showed, he said, that capabilities such as the instant replay of football action on digital video recorders get high use when consumers first encounter them, but that after a short while consumers lose interest and simply view what comes at them.[29] Yet marketing and media executives often said at conferences, in one-on-one conversations, and in the trade press that consumer interest might well change with future generations of viewers. Turn-of-the-century 18–55-year-olds were stuck in an old model of television viewing. Their children, and their children's children, seemed much more comfortable with multitasking and with frequent clicking. An executive at TiVo said: "The issue for us and others [in interactive television] is getting past the inertia of how people watch TV. It hasn't changed . . . our whole lives. What we've learned is that all of the advertisers we've worked with have accepted that sometime in the future, the consumer will be in charge. Once you see that, you see there are more opportunities than barriers."[30]

By 2005, according to some in the television industry, sponsors were pushing program producers to develop commercial-friendly approaches to audience interactivity. "Advertisers are asking us to come to them with these kind of ideas," said a vice president for interactive development at NBC. The requests, he added, created tension between the parties, "because they want us to build it [first], and we want them to commit to advertising."[31] In late 2004, the challenge led the American Film Institute to invite creative teams from television companies to see what they could do with interactivity. According to the *Wall Street Journal*, executives said that "the prototypes they came up with, which go far beyond anything available today, are the types of shows that will soon appear on television screens."[32]

Participants in the AFI meetings approached audience interactivity with programs in two ways. Producers of "reality shows" saw the technology as an opportunity to get viewers directly in touch with products that sponsors had paid to integrate into the programs. The team from NBC's Bravo cable network, for example, figured out how to get viewers to opt to receive recipes and phone messages about products while watching *Queer Eye for the Straight Guy*. The show would indicate when a tip was available for transmission. To accept a tip, one would press certain buttons on one's phone. The tip would then migrate from the television to the phone, so the viewer would be able to consult the tip while shopping.[33]

Producers of music videos and of programs with narrative threads seemed less interested in tying interactivity directly to products than in devising ways to cultivate deeper viewer engagement. Presumably, advertisers would value the engagement because it might bond viewers more strongly to the show and even get them to watch the commercials. Using this logic, the MTV group at the AFI workshop worked to allow its young viewers to use their television remotes to play games superimposed on music videos. "If they're engaged in a game, we would keep them longer," an MTV executive said. An actual implementation of this goal was the Disney Channel's interactive version of *Kim Possible*, a program about a busy high-schooler. In 2005, households that subscribed to the Disney Channel on Demand would be able to hunt for on-screen trading cards with their remote controls and then trade them with friends. To the AFI's director of enhanced television, it made perfect sense that this sort of interactivity would sit well with pre-schoolers. Working with the remote, she said "was so much more intuitive to them than it was to some of their parents."[34]

::

Paralleling work toward interactive TV programming that allow built-in commercials are steps to encourage interactivity with commercials that are separate from the shows. Two approaches stand out. The "targeted pull" approach aims to provide motivated viewers with a place to find commercials they want to watch. The "customized push" tack sends to viewers commercials that appear to be traditional but are really tailored to their background through database analysis. Both these methods can be combined to yield "pull" commercials that are customized. Moreover, current technology makes customized interactive product placement practicable. The question is whether—or more likely, when—marketing and media practitioners will spend the sizeable amount of money needed to roll out some of the more high-tech of these activities.

TiVo—which advertising people saw initially as the "Darth Vader" of television commercials—was an important force behind the "pull" approach. TiVo executives concluded that their company needed revenue from advertisers to survive. TiVo's ability to tinker with the television signals that its million-plus viewers receive also gave it the ability to point them from regular commercials to its Showcase, a space for watching commercials "on demand." For example, the investment firm Charles Schwab & Co. paid TiVo to link a 30-second network spot starring the golfer Phil Mickelson to Showcase via a special symbol on TiVo-attached sets. The symbol signaled to viewers that if they clicked they could see more; it turned out to be a four-minute video about the company and three segments with the golf pro. Viewers of Showcase could also order information from Schwab via the television set. They could then return to the program they were watching at the exact point they were watching.[35] As the director of operations at the agency that oversaw the Schwab presentation noted, this use of the DVR marked a "TV-plus approach" that "gave us great response from the hand raisers, as . . . a direct-response medium."[36]

The idea spread beyond TiVo. In 2004, EchoStar's Dish Network, having linked its set-top DVR to its satellite delivery system, began offering advertisers similar packages to about 9 million subscribers.[37] Another version of this targeting of viewers who might want information on particular products involved video-on-demand (VOD) cable services. VOD refers to programs that are stored digitally on huge servers at a cable company and sent to a subscriber's television via the digital cable box when the subscribe presses a certain remote-control button. The advertising piece involves placing long-form commercials just after VOD programs that seem to harmonize with the interests of particular advertisers. For example, in early 2005, on Comcast cable systems, a VOD program from the Discovery Science Channel was preceded by a message from General Motors that urged viewers to stay tuned at the end of the program for a video about the new Corvette. That 15-minute video was a documentary-style message from GM highlighting the new Corvette's advanced technology—a message that might be interesting to the kind of people who would select a science program. Borrowing from the experience of VBI virtual advertising channels and TiVo Showcase, General Motors also worked with Comcast to develop the GM Showcase, a VOD channel that allowed viewers to select similar programs about other GM products and to request more information. "The thing about selling new cars and trucks is that in any given market, only about 1.5 percent of the population is looking for a new vehicle," said GM's general director of media operations. The VOD world's advantage over linear television, she said, is that it allows marketers to reach those individuals.[38]

For consumers who might not be motivated to pull ads to them, Comcast and other cable operators were experimenting with variations on the traditional "push" approach. The aim was to link database-marketing capabilities to 30-second commercials during the programs. In one sense it was still traditional, because it assumed a "lean back" consumer who didn't have to change behavior in the face of ads. At the same time, it marked the drive toward digital marketing discrimination in the television world. Comcast promoted its ability to send different commercials to different areas on the basis of distinctions that it and its advertisers found between those areas. With a service it called Adtag, a car dealer could add different voiceovers to an advertisement for different geographic locations. Adtag allows advertisers to run the same 25-second commercial throughout a market, then finishes the spot with customized 5-second "tags" that give specific information to the appropriate geographic location within the market. In the words of Comcast Spotlight's website: "Tag #1 could say 'Visit our new showroom on Main Street!' and tag #2 could be 'Visit our showroom on Route 78 today!'"[39]

Comcast suggested that its Adcopy service offered even more startling change:

Picture this: You're at home one evening watching ESPN's SportsCenter. At precisely 7:09 pm, a commercial comes on for the new Chevy Blazer. At the very same time, a friend of yours is also watching SportsCenter; only instead of seeing a Chevy Blazer commercial, he sees a commercial for the all-new Chevy Equinox.

Chevrolet planned it that way. They utilized Comcast Spotlight's targeted TV application Adcopy, a market segmentation product that broadcast networks simply cannot offer.

From one point of view, what Comcast was offering advertisers was nothing more than the ability to use the distribution equipment (the "head-ends") of Comcast's local systems to target zones.[40] Because the systems covered fewer homes than broadcast signals, advertisers could discriminate between smaller areas based on data from companies such as Claritas that provide information about wealth, lifestyle, and purchasing habits based on ZIP codes. For example, a dealer of upscale cars might want only to reach the zones in a cable company's systems where people above a certain income reside. Using its interconnects between cable systems it owns, the cable company could send the appropriate commercial to its different systems, which would then send the commercial to only those head-ends that fit the income parameters.

Daimler-Chrysler used Comcast's capability to aim commercials featuring the high-priced Chrysler 300 and other commercials featuring lower-priced vehicles to different zones in Pennsylvania. Similarly, Ford dealers in New Jersey and in parts of New York ran nine customized cable ads for trucks, offering different deals depending on geography, the income of the area, and other demographic data. Higher-income areas such as Princeton saw ads with a discount for the top-of-the-line F-150 pickup called the Lariat. Lease-oriented ads for the basic model were delivered to lower-income areas.[41] This kind of targeting encouraged greater differentiation among neighborhoods by television advertisers than was previously practicable. The Visible World technology that powered it, though, had far greater capability. Backed by such huge advertising players as WPP Group and Grey Global, Visible World's Intellispot system could create and deliver television commercials that changed messages and creative elements in real time.[42] It did that by creating different layers for parts of the commercial that could be changed digitally. At cable systems' head-ends, a commercial was placed on servers jointly run by the cable firm and Visible World. The layered nature of the commercials allowed the advertisers to change the message for that zone on the basis of anything from the weather to the time of day to the day of the week without delivering multiple tapes. A Visible World executive said that people in homes receiving the commercials would not know (unless told) that they were getting messages targeting to their area. He added that the customization could easily be combined with interactivity. Visible World could work with interactive advertising firms to allow people who received the customized commercial to click on elements of the commercial to learn more or request information.[43]

The Intellispot system presented the prospect of even greater customization. The software could implement thousands of versions of a commercial in seconds by changing features from music to voiceover to characters to graphics. In the case of a car commercial, for example, one layer might involve the vehicle; another, the driver; a third, the kind of highway; yet another, the song played in the background. Using database instructions to software in household set-top boxes, Visible World could create commercials customized to different individual homes, not just head-end zones. For example, a firm could seamlessly send an African-American household a car commercial with an African-American female driver while it sent a Korean-American household the same commercial with a different price incentive and a Korean-American male driver.

Although it was technically quite feasible for cable systems to implement household-customized commercials in 2005, that was happening only in scattered tests. One reason was relatively sparse use of the digital set-top boxes needed to process the commercials. Gerrit Niemeijer, Visible World's chief technology officer, noted in 2005 that cable firms were loath to spend the $1 or more per box that would be required to give a digital box the ability to process his firm's layered commercial. One dollar seemed like a small amount, he pointed out, but for a multiple system operator it added up to substantial money; Comcast, for example, now has more than 20 million subscribers.[44]

And Niemeijer mentioned an additional hurdle: privacy concerns. He said he had heard cable-system executives express worries that the kinds of personal-information issues that swirled about the web would hit them. He said he understood their hesitation. "Cable firms are gun shy," he said, but they are also practical. "Cable firms know there is an enormous value to be had. It's very simple: more than $60 billion in advertising money is spent yearly in television. Major advertisers would like to change spending habits if they can. Cable companies know this, and know they have the ability to do it. They say [privacy] is a problem. That just means they will be careful about that."[45] Niemeijer firmly expressed the opinion that the Cable Television Act of 1984 doesn't prohibit cable firms from sending customized commercials to households. No one seems to contest that point, despite the complexity of the act's privacy section. Using the same kinds of tortured clauses and possible escape hatches common to corporate privacy policies, the section seems to first take away and then return to cable firms the right to give marketers ways to discriminate among subscribers. The section seems also to say that cable systems cannot sell personally identifiable information to marketers—except that they can sell basic "mailing list" information. That means the names and addresses of individual subscribers. The section also gives cable systems the right to collect a lot more data about subscribers for their own uses if subscribers give "prior written or electronic consent"—or if it is "necessary to render a cable service or other service provided by the cable operator to the subscriber." One such "other service" may well be advertising.[46]

The Cable TV Act is also ambiguous when it comes to using data in which the names and locations of individual subscribers are masked. The act notes that its prohibition on using personally identifiable information "does not include [i.e., refer to] any record of aggregate data which does not identify particular persons."[47] This formulation makes it unclear whether a record of aggregate data may include specific information about a particular household collected by the cable firm but stripped of names and locations.

Cable firms seemed to interpret their rights to use customer data to the limits of the Cable TV Act. On its Spotlight website, which is meant for advertisers, Comcast implied that it was collecting personally identifiable data about its customers and then turning the data into aggregate information by head-end zone for its advertisers. Spotlight, it said, "provides advertisers with sophisticated research to maximize the effectiveness of Adtag and Adcopy by targeting viewers based on aggregate geographic, demographic, psychographic or other characteristics of the consumers residing within specific areas."[48] Cable firms also read the cable law as allowing them to go farther with personally identifiable information if they glean it from publicly available databases. Time Warner Cable was quite explicit on that point in its privacy policy. A cable operator, it said, "may add to its mailing list publicly available information about subscribers that is obtained from third parties."[49] Comcast was more circumspect. In a privacy policy that mirrored the opaqueness of the privacy section of the Cable TV Act, it suggested obliquely that it offers third-party data to its advertisers along with their names and addresses.

Spotlight's managing director Hank Oster stated flatly that privacy laws prevent his Comcast division from selling subscriber names and addresses to advertisers. Spotlight does collect loads of demographic and viewing information about households in head-end zones and then aggregates the data in order to interest sponsors. Spotlight will also take data from individual advertisers about individual addresses that they want to target and confirm the percentage of the zone's households they represent. For example, the cable advertising marketer will confirm to Kraft that 25 percent of an area's cable homes are addresses that Kraft knows buy the company's cheeses. That high percentage might encourage Kraft to advertise in that Comcast zone, or purchase Showcase programming, because it is higher than the national average. Oster went on to say that interactive commercials on Spotlight could send interested customers to a website where customers could ask for more data and even be tracked. He recognized the irony of enticing consumers to the less stringent privacy regime of the web at a time when all the technologies were melding from a marketing standpoint. He simply said that different rules had developed for different media, and he was abiding by his. He insisted that privacy laws would not allow him to do it even if the server technology allowed the customer data to be shielded from the advertiser or from Comcast.

At this point, household-level commercial targeting is technically impractible for Spotlight. Oster's pitch to advertisers who might be interested in such targeting is that they would be overwhelmed if it were practicable: "How do you build a media plan on millions of households? No [media buying firm] would have a stewardship plan whereby they would know how to determine the individual households or to invoice or track the activities. Nobody has done that before." Oster acknowledged that household-level targeting might some day happen with television but he advocated what he called a "crawl-walk-run" scenario. He noted that advertising agencies and their clients were pushing to get to individual homes and even people without realizing the complexities of such activities. "Advertisers have to learn how to use television in a targeted environment before they get to the individual," he insisted.[50]

Executives at Visible World disagreed. As they saw it, commercials tailored to individual homes awaited only the cable firms' desire to implement the technology. Niemeijer, the chief technology officer, interpreted the Cable TV Act as allowing it to send a commercial customized to what the cable system or advertiser knows about the people in a dwelling. He said it was legal as long as the information is public or the party not owning the personally identifiable information—the cable system or advertiser—doesn't have a chance to see it.[51] Another Visible World executive added that often the advertiser is more active than the cable firm in bringing substantial household data to the marketing situation. In that event, the advertiser would download personal information to the cable box that would help create the custom layers for the target. The information would disappear after the commercial was created. Because neither the cable firm nor Visible World would share that information, it was all quite legal and potentially very powerful for sending different commercials offers to households—and possibly eventually even people in those households—on the basis of what marketers conclude about them.

"This is the future of TV advertising," contended a venture capitalist with an investment in Navic Networks, a startup firm competing with Visible World. "If I were to factor what TV advertising may be like in three to five years, I think today's concept of producing blanket TV ads will be analogous to dropping leaflets out of an airplane."[52] Backing up his claim, Forrester Research had found a high percentage of database-marketing executives for major financial, telecom, and retail firms very interested in household-level targeting of television advertisements. Their desire to reach the right people was so high that they said they would pay between 50¢ and 60¢ for each ad delivered to a household. It means spending as much as $600 to reach a thousand viewers at a time when conventional prime-time television charged between $30 and $50 per thousand. "That's off the chart," exclaimed a Forrester analyst.[53]

Visible World was, in fact, working to make that idea attractive by making it easy for advertisers to find the households they wanted. The firm turned to Teradata, an Ohio-based company, owned by NCR, that sells consumer behavior data to advertisers such as Travelocity. The advertisers were using it to target consumers with tailor-made ads on the internet. The idea was to do for television what it does online: deliver television ads that are not just tailor-made to people who live in certain neighborhoods, but are also tailored to individual interests. "Our end-game is to mass customize commercials as granular as you can get," a Visible World executive told *Advertising Age* in connection with its Teradata project. "We envision the day—in three to five years—when consumers will actually request commercials [customized for them], and that is the ultimate relevance."[54]

That commercials can be made interactive and tailored to households and even to individuals is such an attractive idea to marketing and new-media practitioners that they often discuss it as the ultimate antidote to ad-skipping. Their expectation is that viewers will pay attention when people see and hear products and claims that speak directly to their interests. An obvious addition to this armamentarium is customized product integration directly into programs.

::

Asked about the possibilities of customized product placement, executives from marketing and technology firms replied that, although practicable, it would be even harder to implement than customized commercials. Gene Dwyer, Director of Technology for Princeton Video Imaging, said that his company has the capability to custom-insert products into shows during real time. Princeton Video Imaging has for several years been digitally inserting products into reruns of *Seinfeld* and into other syndicated pro-grams.[55] The firm also incorporates advertising that only viewers see into major league baseball and soccer stadiums. Those insertions take place in real time. For instance, Princeton Video Imaging places virtual advertise-ments behind home plate during Major League Baseball games.[56] Dwyer noted that at present the desire to customize these activities surrenders to the demands that household-level customization makes on the set-top box. Database-driven product placement requires even more computing power in the digital set-top box than does the Visible World–type commer-cial creation. Creating a 30-second commercial requires the combination of layers. Customized product placement requires that plus integration of the material into the flow of ongoing programming. A Visible World exec-utive said that his firm's technology could be adapted to create customized placements in programs so that homes would receive programs with differ-ent props in the scenes on the basis of what marketers know about, and want from, the households. Yet Visible World has decided not to pursue that route, because the demands would divert the firm's attention from its primary mandate, which is to spread advertisers' ability to create and dis-tribute commercials in many versions in real time on as many platforms and with as much interactivity as desired. Startup companies have floun-dered by pursuing missions that are too broad, he said.[57]

It is, however, a pretty sure bet that within 15 years customization of all sorts of commercial messages will be feasible and competitively essential. At this point, the biggest logjam is technical: There are not enough digital set-top boxes in the approximately 70 percent of American homes that get their television via cable, and the boxes that do exist are too primitive to accommodate real-time customization. But the situation is very fluid. Cable firms already see strong reasons to pepper their subscribers' homes with digital set-top boxes and add substantial computing power to them. The particular motivation is strong competition from satellite and phone companies ("telcos") that aim to compete with cable firms to provide a "triple play" of voice, video, and data. Consumer electronics companies such as Apple, Hewlett-Packard, Sony, and Philips also stand ready to com-pete with all these firms in the "home entertainment" space.

The national rollout of digital television by large phone companies such as Verizon using internet protocol technology will increase the cable firms' need to deploy smarter interactive capabilities. Ed Graczyk, marketing director at Microsoft TV, Verizon's technology supplier, noted in 2005 that all subscribers would get HDTV, DVR, VOD, and an interactive electronic program guide. He added that Verizon and various third-party vendors will, over time, easily add various advanced internet-based interactive services to the offering. The technologies are likely to allow viewers to personalize the program guide and to conduct programming searches and schedule recordings from personal computers and other connected devices. Also in the wings are applications that will incorporate internet content and personal media into the viewing experience, allowing viewers to use integrated entertainment and communications services that will work with set-top boxes, with personal computers, with wireless phones, with other hand-held devices, and with the Xbox gaming platform.[58] It isn't difficult to see that such features offered by their competitors will lead cable firms to respond in the direction of converting their analog cable customers to digital. According to Graczyk, Microsoft believes that the cable industry will also eventually adopt technology that streams television programs over the internet—so-called internet protocol television (IPTV). His persuasive contention that the cable industry will have to eventually move in an all-digital direction is worth quoting at length from an interview in *Tracy Swedlow's itvt newsletter:*

> The cable industry has talked quite a bit about 'the all-digital cable network' and 'the next-generation network architecture,' and I think it is pretty clear from what they are saying that IPTV is the direction that they see their services eventually going in. One of the huge advantages of an IPTV environment is that your TV infrastructure is essentially the same as the infrastructure used to deliver all your other services—unlike the situation today in cable for example, where the TV delivery infrastructure is substantially different than the broadband infrastructure. In fact, sometimes when you order TV and data service from your cable operator, two different engineers come to the house, one to install the digital cable service, one to install broadband. With IPTV, instead of sitting on this proprietary architecture, the TV becomes a member of your eco-system of devices in the home, capable of communicating with PC's, gaming consoles and all the other devices people rely on in their day-to-day lives. Besides, not only is IPTV in many ways a generation of technology ahead of what cable operators are using for their video infrastructure today, but they already have a great IP network and huge broadband penetration, thanks to their data services.[59]

Graczyk added that one reason the cable industry is not already moving aggressively to IPTV is "because they've got a huge legacy investment: they've got millions of subscribers with millions of set-top boxes and all their other proprietary infrastructure in place. They've invested a lot in those networks, and I think Wall Street continues to look to them to generate all the return they can on these existing investments. Because, even though they've got the core building blocks in place to roll out IPTV, it's still going to require substantial additional investment."[60]

With tough competition and high technology costs, profit margins will undoubtedly be narrow down through the producer-network-cable/telco/satellite food chain. It stands to reason that every party involved will applaud looking for ways to add a stream of advertising revenue to the mix. Josh Bernoff may not have been wrong when he wrote his "smarter television" scenario; he may just have been ten years early. In 2005, companies with techniques for integrating interactive and even database-driven selling into television-like programming were streaming into advertising agencies and media firms. Rob Buchner of the Fallon Worldwide advertising agency spoke of getting a headache thinking of the various options that he had to choose for his clients. Buchner and his Fallon team had gained some fame in the advertising business in 2003, when short action films by famous directors highlighting BMW automobiles received millions of downloads. Fallon followed that with Amazon Theater narrative films on the shopping site's home page that included objects in the movies that could be bought through Amazon. Not only did these films also yield millions of downloads; they were especially distinctive for posting links at the end of each movie that led to pages for buying various items.

After Amazon Theater's release in late 2004, a group from the broadband media software firm Maven Networks approached Rob Buchner with an idea that took the sales idea even further. Maven could create a version of the downloaded Amazon presentation that would have all of the products not in links at the end but as a stream of objects at the bottom of the screen. The viewer could turn the stream on or off, reflecting the emerging view that product integration should be subtle. If the viewer decided to keep it turned on, he or she could click on objects to learn more and even purchase them while continuing to watch the movie or stopping it. Buchner, who was wowed by the possibility, emphasized that a person downloading the Maven film to a laptop computer could view it and the product stream information anywhere, even during an airplane flight. Only the purchase would need a web connection. Of course, Maven's application was designed for use on the internet, not on a traditional television set. Yet with "television" becoming a fully digital phenomenon over the next decade or so, and especially with the rise of IPTV, many distinctions between the capabilities of internet and television technology were bound to disappear. Bringing the techniques to the large home screen would be a natural next step. As a General Motors marketing executive put it when she compared viewers' use of Comcast's advertising Showcase to the internet: "The way we see this shaping up is that television and the computer are going to merge into one entity."[61]

In the early 2000s, advertising agencies had created commercials on the web that aimed to pull millions in the target audience toward the sponsor's website. Some had become cultural events of sorts. The action-filled BMW ads stood out. So did the hilarious "Adventures of Seinfeld and Superman," created for American Express by Ogilvy's New York office.[62] In the short spots, the comedian Jerry Seinfeld bantered and bickered with an animated Superman, who was not able to help his friend as much as the American Express card could. Reflecting on these ads-as-magnets, *Adweek* commented: "In good conscience . . . many [marketers] can't justify throwing big money toward producing content for these platforms when the penetration levels [that is, the visits to the sites] are still fractional."[63] Yet Rob Buchner of Fallon Worldwide and Mark Sitley of Euro RSCG noted at a 2005 conference that the client-sponsors of such unusual commercials tended to be extremely happy with results, which often brought millions of clicks, encouraged downloads, stirred buzz in various media, and demonstrably led those in the target audience to ask for more information about the product.[64] Buchner remarked that the BMW films were "entertaining relief for the target audience not watching TV," that 10 percent of the tens of millions of viewers of the BMW films downloaded them, and that 28 percent asked for more information about the car. Sitley further highlighted the direct-marketing payback from such commercials when he said "leads are worth gold." James Warner, Executive Vice President of Avenue A/Razorfish, summed up the collective sense of what these and other alternative forms of audio-visual advertising portend. "We are," he said, "training consumers to experience ads in new ways."[65] The panel did not talk about using databases to selectively display ads. In a later conversation, though, Buchner noted that his team at Fallon was seriously interested in the possibility. The panel moderator, he pointed out, asked the members to describe changes that would take place in the next year. Mirroring the executives at Visible World and Princeton Video Imaging, he saw it as a few years down the road.

One doesn't have to look far, however, to find attempts to wed the web to television in ways that serve as a model for the database-driven commercialization of television. Consider Sony's use of GSN, its cable/satellite network devoted to gaming, for an interactive television loyalty program. Formerly called The Game Show Network and a haven for reruns, the channel moved in 2005 to reposition itself as an interactive gaming hub that paraded Sony's gaming technologies. Because only Time Warner's Hawaii system allowed viewers to interact with the games on the television set, though, GSN encouraged viewer interactivity through a "two-screen" approach: players went on the web to interact with the games they saw on the television. The goal of the loyalty program was to help improve the image of the GSN network among "younger viewers" as well as to increase the number of people tuning into the channel and length of time they spent with it. Dubbed GSN Rewards Powered by My Sony, the program gave viewers who interacted with the television games on the GSN website the chance to earn points from the My Sony loyalty program of the network's parent company. Viewers received 1,000 points for signing up, all points were doubled during their first membership month, and then they received 125 points each time they went to a web version of game which was synchronously running on the television network. "With the loyalty program, you don't have to get lucky, and you don't have to be the best player," said Dena Kaplan, GSN's senior vice president of marketing. "You just have to watch and play along, and you will earn rewards." She noted that web players also would get points for clicking on website ads. "Rewards members are going to have a better affinity for the advertiser, because they know that every time they engage with their ad, they're earning points towards something tangible."

The attempt to cement relationships also included a permission marketing component: GSN Rewards members were able to opt into a program that would target them by their ZIP codes. That would lead to offers via advertising banners or email, inviting them to purchase broadband internet access and other products and services from their local cable operator; they received points for clicking on those offers, too. And there was more: GSN encouraged WebTV viewers to redeem points for "Cable Cash" (a promotional program run by the American cable industry) that they could use to pay their cable bills or purchase VOD and pay-per-view movies.

A database-marketing component made the loyalty program especially valuable. GSN and Sony promoted the new program to members in their respective databases and shared all customer information generated by the program. In February 2005 GSN had a database of more than 2 million viewers who had registered to play its ITV applications, while the My Sony program had about 3 million members, acquired through "Sony Music Rewards" and other programs. GSN Rewards members and My Sony members received "GSN/My Sony," a weekly online newsletter featuring links to special offers from the companies, and ads from GSN sponsors (which also awarded points to viewers who clicked on them). Kaplan emphasized the ability of the databases to help both firms, and the special utility of the My Sony database to GSN advertisers. "The stats on the My Sony members are that they are young, tech savvy, and love entertainment—which of course makes them very desirable to our sponsors, she said."[66]

::

Just a bit into the twenty-first century, then, advertising and media practitioners see "television" very much from the standpoint of the process of database marketing that already has begun to emerge on the internet. They know that the technology is not yet advanced enough to combine interactivity, targeted tracking, data mining and the cultivation of relationships in one advertising application. They are, however, testing all aspects of these activities with the sense that if they don't understand new models, their competitors will. The new perspective that is emerging would have seemed hardly plausible to the medium's gurus only 20 years ago, when audience "tonnage" was still the dominant coin of the realm. Now network personnel who still often sell tonnage—for example, the salespeople at ABC, CBS, and NBC—increasingly have to face advertising people at industry conferences who question the long-term viability of their business model. A new language of television strategy is evolving in tandem with targeting and customization tools. Stuart D'Rozario, group creative director at Fallon Worldwide, says that targeted ads make what has historically been an impersonal mass medium personal. Hank Oster, managing director at Comcast Spotlight, adds: "It allows you to connect at a more specific level and more personal level with people. You don't need to speak to the masses if you can speak to those few and you have the content that is relevant to that person."[67]

Oster did indicate a privacy concern about helping marketers "drill down" to individual households. In any event, his division is at this point technically unable to accomplish household targeting of customized commercials, let alone programs. In view of the way web-based media firms are exploiting audience data for advertising, it seems plausible that Comcast's perspective on household information will move in that direction as competition with satellite and telephone providers of television signals generates strong pressures for cable firms to increase revenues through advertising. The Spotlight division will likely then ramp up its technologies and redefine its rules to allow advertisers to configure materials for individual households, or types of households, that opt in or do not opt out for these activities. Vendors of specialized business software for cable, satellite, and telephone operators appear to believe this will happen soon. They are creating database programs that encourage the kinds of customer-relationship-management systems that are becoming the norm on the web. The software allows the firms that sell the triple play of television, internet, and digital home phone service to bring what they learn about customer activities on each medium into integrated databases. That will allow the firms and their advertisers to discriminate among customers. They will select the best households or individuals to receive particular ads, discounts, and programs on television, on the web, and even through the telephone.

As one business software provider expressed it, the idea is to "turn every customer touch point—every point of presence—into a point of sale," using "records on all clients that can be tracked and used with a lot more sophistication than service providers are currently doing it with."[68] In its triple-play form, and especially among cable, satellite, and the Baby Bell telephone companies, the strategy is most often discussed in terms of the home and office. Marketers are applying these database marketing techniques outside the home and the office, too.

6 :: The Customized Store

My fantasy is that our best customers will have loyalty cards, and when they walk into a store the manager will be alerted. We want to know, for example, if a customer was on our Web site looking for plasma and LCD [liquid-crystal display] TV sets. Maybe that customer didn't buy anything but has an interest. We could create a tipping point for the customer. Maybe we put a personalized coupon while [he or she is] in the store.

Someday, if a customer makes a big transaction, we could have our CEO W. Alan McCollough's BlackBerry go off. The customer would get a phone call from McCollough or another top-level executive, who would thank the customer for his purchase and ask how the experience was.

—"Circuit City's Fix-It Time," *Business Week Online*, January 20, 2005

These comments were made by Michael Jones, chief information officer for Circuit City, American's ailing number-two consumer-electronics retailer. Though Jones called his notion a fantasy, other companies were already describing it as reality:

- The *Wall Street Journal* reported that Circuit City's main competitor, Best Buy, was using its rich databases to divide its customers into segments and was teaching its salespeople how to identify "better customers" on the basis of their shopping behavior. Best Buy was also initiating a strategy to discourage the "wrong" customers from shopping at its stores.[1]

- In December 2004, the manager of information systems at Dorothy Lane Markets, a mid-size Ohio-based supermarket chain, described its strategy as "top-customer centered." In a presentation at the Global Electronic Marketing conference, Amy Brinkmoeller outlined how Dorothy Lane segmented customer data from its Club DLM loyalty program into various levels, and how it gave increasing amounts of personal service to customers with increasing purchase levels.[2] By discontinuing weekly promotions, she said, Dorothy Lane Markets "essentially fired 1,500 customers" who came to the stores hunting for bargains.[3]

• In December 2004, the marketing vice president of Leading Hotels of the World told a reporter for Direct magazine that the chain had made its Leaders Club loyalty program "by invitation only." The move, he said, was intended to give the a special sense of worth to the 90,000 members, who paid from $350 to $600 for a night's stay at a Leading Hotel. "The thing that appealed to members was the exclusivity," he said. "They saw that as a form of personal recognition." Asked if he has a profile of these "best customers," he reeled off numbers from a database: "We know 50 percent are C-level executives, like CEOs, CIOs, CFOs, or COOs. The average household income is $240,000 plus. Average net worth is $1.7 million. Eighty-four percent are college grads and 60-odd percent have done postgraduate study. The critical age group was 45 to 50 years old. They take 24 trips a year—15 business, nine leisure—and spend 63 nights a year in luxury hotels. Seventy-two percent are men, but we want to see a more equal ratio of men to women."[4]

These are not isolated cases. Major developments in the use of database marketing at the retail level are paralleling the developments in digital media that I described in previous chapters. With new information technologies, new analytical techniques, and a changing commercial environment, stores are thinking about and treating customers in new ways. Like the new media regime, this new regime is built on data mining, segmentation, targeted tracking, interactivity, mass customization, and the cultivation of relationships.

The customized sales environment that Michael Jones dreams about is profoundly different from the one that advertisers were urging stereotypical Joneses to keep up with during much of the twentieth century. A new message about being a consumer is just beginning to percolate through American society. This message is that, in order to get the best treatment, customers must enter into new bargains in various places where they shop. If they buy the right amounts of goods at the right prices, and if they provide the right data, the retailer will reward them with announcements and deals that they will enjoy and that will encourage further interaction and loyalty. If they don't fit the profile and provide the data, they will not count nearly as much as those who do. They may be charged more than "good customers," they may not feel quite welcome, and they may even be nudged away.

The methods used to announce and to implement such bargains have been developed most highly by firms (such as Leading Hotels of the World) that target as customers wealthy individuals and businesspeople whose spending is reimbursed by their employers. During the past few years, though, less upscale businesses, including supermarkets, have accepted the industrial logic of marketing discrimination and have taken great strides toward implementing it according to their own needs. These developments support and are supported by major changes in digital media. Retailers and their suppliers are learning to use the internet, interactive television, mobile phones, and other media to find new customers, to gather information on new and old ones, and reach out to customers with ads and content rewards that are increasingly tailored to match what the databases know. They aim to create customized environmental surrounds that inspire trust and return business. The demand for data to make the most of the new bargain, though, inevitably means that marketers may be exploiting information about their customers in more and different ways than the customers expect.

::

When asked to reflect on the growth of database marketing, retail executives and consultants often associate the collection of information about consumers with the ways storekeepers of old connected with their customers. In the days before impersonal chain stores, they say, store owners and clerks kept mental tabs on their customers' attributes and habits. Writings about that era sometimes reinforce this picture. In an essay on her visits to the classic New York department stores of the mid 1940s as a college student, Letitia Baldrige recalled: "The salespeople never forgot you. They made little notes on you, your family, and where you were in life each time you stopped to buy or to custom order their merchandise. It was such a safe, predictable world. It was also intensely personal—everything directed at you and no one else."[5] There is, of course, no way to know how much these memories are romanticized through the miasma of time. Quite likely, in small towns and neighborhoods shopkeepers and clerks did, and still do, get to know their customers and their situations well. When it happens, it is a classic case of personalization in the service of commerce—that is, of direct and ongoing human interaction within a sales situation. Database marketers try to mimic that. They do it through mass customization, which involves using media to send customers messages that have been chosen or assembled by computers from information on them that the firm has gathered. By sending material that matches an individual's interests, the firm hopes to get that individual's attention and to affect his or her purchasing behavior.

As I noted in chapter 3, databases are not new. Marketers have been buying lists for more than 100 years. The use of computers to categorize consumers became common in the 1970s, particularly for direct marketing. In the 1980s and the 1990s, as database-driven loyalty marketing emerged, financial and leisure firms and elite retailers began to adopt the logic of segmenting their customers and pursuing the 20 percent who supposedly accounted for about 80 percent of sales. Though it wasn't fully established even in those sectors by 2000, the industrial logic of investing in database marketing in the service of customer-relationship management was catching on. At first, such thinking didn't have much of an effect in the broad retail sector (the realm of mid-price department stores, drug stores, and supermarket chains). Then that began to change, and rather quickly. The change was due in large part to the huge retail chain Wal-Mart.

In one sense, the name Wal-Mart stands for a *kind* of store: a supercenter that sells food, drugs, and general merchandise. But Wal-Mart is so strong in this category, and its shadow is so large in retailing overall, that executives and consultants talk about the company far more than the store type. Wal-Mart's effect on merchandising strategies is enormous. Most Americans barely knew about Wal-Mart in 1985, when it was focused on rural and small-town markets and on selling goods made in the United States.[6] Today, however, Wal-Mart is the world's largest retailer. Employing 1.5 million, it owns more than 3,500 discount stores, supercenters, Neighborhood Markets grocery stores, and Sam's Club membership warehouse stores. Its revenues in 2004 amounted to $288 billion—about four times those of Home Depot, the second-largest retailer. It sold more goods than Target, Kmart, J. C. Penny, Safeway, and Kroger combined.[7]

One analyst's conclusion—that Wal-Mart had "become a steamroller that seems to be able to take on any and all competitors"[8]—was certainly shared by manufacturing and retail executives in the 2000s. Wal-Mart had become the largest retailer of toys and CDs and the number-two grocery retailer. After less than 20 years in the grocery arena, it had captured nearly 20 percent of the business.[9] At marketing conferences and in retail trade magazines, Wal-Mart's accomplishments were a routine topic of fear and awe. A noted retail consultant considered competition from Wal-Mart a major factor in the 2005 Chapter 11 bankruptcy of the southeastern grocery chain Winn-Dixie.[10] "Wal-Mart's success in grocery retailing has been staggering," *Chain Drug Review* noted in 2004.[11]

Wal-Mart is secretive about its use of databases. Competitors often try to make inferences about its information strategy from comments made by its executives and statements made by its suppliers. Analysts typically note that Wal-Mart's huge investment in databases plays a large part in its general retailing success. The company's slogan is "Every Day Low Prices" ("EDLP"). What this means is that its selling strategy focuses on being known for offering a product at the same low price all year round. When setting that price, Wal-Mart tries to anticipate the lowest price its competitors will charge at different times in the year. Wal-Mart does not use coupons or loyalty programs.[12]

Analysts agree that Wal-Mart has long used information about the movement of merchandise through its system to keep costs down. It has done that by collaborating and sharing information with suppliers so that they take more responsibility for their goods than they would with other retailers. Central to that practice is Retail Link, a powerful tool first developed in 1991 to facilitate the sharing of information between Wal-Mart and its vendors. Retail Link's database contains years of sales information for every product sold in any Wal-Mart store. It lets suppliers see how merchandise has sold historically and how it is selling currently in any of the stores. Using the data, Wal-Mart meets with each of its suppliers to establish sales goals for the coming year. The chain is well known for its toughness. According to the *New York Times*, "a manufacturer that fails to meet its sales target—or has data-documented problems with orders, delivery, restocking or returns—can expect even tougher negotiations in the future from Wal-Mart."[13]

Analysts also point to Wal-Mart's dedication to making its supply chain hyper-efficient. This begins with the choice of manufacturers—mostly with factories in China[14]—that can reliably provide products at the lowest possible price. It continues with working to reduce the costs of transporting, warehousing, and shelving goods. That includes forcing suppliers to adopt the Global Trade Identification Number system for individual items so Wal-Mart can use electronic product codes to synchronize the identification of all products in all store and warehouse computers. It includes requiring suppliers to tag shipments to some of the retailer's distribution centers with tiny radio-frequency identification (RFID) devices that will allow it to track the movement of goods to and from the distribution centers and at the stores' loading docks.[15] (Eventually, RFID tags on individual items could help track each piece on the shelf.) It even involves requiring many suppliers to own the inventory held in Wal-Mart warehouses until the product is sold—a process called *scan-based transactions*. That means that Wal-Mart does not pay for the item until it is scanned at the checkout—a burden on the supplier, but a boon to the retailer.[16]

Checkout scanners send information to computers that can store information about individual product sales as well as what individuals purchase (as determined by names on credit cards or on driver's licenses shown when cashing checks). The amount of data collected is extraordinary. In 2004, Wal-Mart reported having 460 terabytes of data stored on mainframes at its Bentonville headquarters—far more data than made up the entire internet at the time, according to experts.[17]

Gib Carey, who led an effort by the consultancy Bain & Company to understand Wal-Mart, asserted on the basis of his team's analysis of prices that Wal-Mart's general objective is to price each item 25 percent below its competitors' prices. Carey estimated that about 20 percent of that difference could be attributed to Wal-Mart's efficiency, about 3 percent to volume discounts that all major chains could get, and about 2 percent to Wal-Mart's getting special price concessions from individual suppliers.[18]

Observers of Wal-Mart say that its goal of low prices reflects its understanding of its core audience. According to Edward Fox, a consultant and a professor of marketing at Southern Methodist University, although "everybody" may shop at Wal-Mart at one time or another, the company perceives its core shoppers to be people of relatively low socioeconomic condition who are too busy, because of work and family, to shop around.[19] Gib Carey saw it somewhat differently, noting that Wal-Mart dominates the consumer niche—about 30 percent of the population, he estimated— that makes purchasing decisions by going to a place that sells what they need for the lowest price. Carey and Fox agreed that Wal-Mart believes that its core consumers trust it to give them the best prices.

Wal-Mart does not focus on individuals, though. Except in the case of its Sam's Club warehouse chain (which tracks the purchases of its many small-business-owner customers in order to show them how much they save), Wal-Mart does not keep data on individual shoppers' purchases beyond the day of purchase. Wal-Mart's lack of a loyalty program means that it does not reach out differently to different slices of its customer base. "Me knowing what you specifically buy is not necessarily going to help me get the right merchandise into the store," the company's head of information systems noted. "Knowing collectively what goes into one shopping cart together tells us a lot more."

Even without knowing the names of its customers, Wal-Mart can conduct analyses on the billions of shopping baskets coming out of checkouts to gauge what they buy, where, and when. It uses basket-level analysis to assess statistically what products at what prices will bring other more profitably priced materials along for the ride. By some accounts, Wal-Mart is particularly aggressive in the market-basket analysis it conducts when it decides to enter a community. Using syndicated checkout data from A. C. Nielsen and from Information Resources, Inc. (IRI), it looks at the items that move in that community. The firm also considers the "complementarity" of items—what sells along with what. Based on this understanding and on auditing of local prices, the chain matches or undercuts the local prices of fast-moving goods. Keeping the prices of these items low by ruthlessly keeping its supply chain "lean," Wal-Mart typically sets the prices for a marketing area—and frightens retailers that sell the same products or similar ones.[20] "Our clients cannot grow without finding a way to be successful with Wal-Mart," said Gib Carey of Bain & Company.[21]

Department stores, grocery stores, and drug stores perceive a need to increase efficiency by squeezing costs from the supply chain through initiatives that mimic Wal-Mart's. Many retailers feel that they cannot compete with Wal-Mart on price, yet it is difficult for many retailers not to claim to do so. That bind results from the perception that overall consumer loyalty to stores is extremely low. Less than half of the consumers of apparel and of office supplies consider themselves loyal to a particular retailer.[22] Consumers seem to be more loyal to mass merchants and to discount clubs, probably because of price perceptions. The sense among retailers is that if Wal-Mart or any other store prices an item lower, their customers will buy that item there.

::

Unable to compete on price with Wal-Mart and similar discounters, many retailers have been searching for the best strategies for winning and keeping good customers. Some consultants suggest that the answer lies in adapting to the varied needs of an area, in terms of the right quality, convenient locations, and variety of offerings, better than Wal-Mart can. Gib Carey argues that the most profitable supermarkets, including Publix and Kroger markets, have tended not to suffer when Wal-Mart comes into a town. Rather, lower-tier competitors that had attracted customers mainly through claims of rock-bottom prices are badly hit. No longer able to claim that they have the lowest prices, they have difficulty surviving, since customers cannot see the benefits of their location, quality, and variety in comparison to Wal-Mart. (This was one of the issues that bedeviled Winn-Dixie when it filed for bankruptcy protection.) Top-tier stores survive and sometimes even grow in the face of Wal-Mart because their reputations rest on providing an enjoyable shopping environment close to many people's homes and on having a range of goods that many customers like.

According to a different stream of analysis, Wal-Mart's long-term weakness is that it has difficulty getting close to individual customers and small niches. This view emphasizes that, with the exception of its Sam's Club wholesale setup, the company does not keep track of individual customers' purchases or reach out to them in unique ways. An important competitive advantage in the Wal-Mart age, then, is to know and reward profitable customers better than Wal-Mart or any other competitors can. Analytics firms with expertise in finding patterns in purchase data are urging retailers to examine the spending behaviors of individual customers. One aim is to develop profiles of "best" or at least "good" customers so as to focus on wooing them. A related aim is to encourage purchases through a better understanding of customers' buying habits. Both goals are changing the American shopping experience.

But before a company can develop a profile of its best customers, it has to know what the term "best customers" means. The answer is not obvious. Consultants disagree on what aspects of a person's buying trajectory make that person most valuable. Some marketers use "past customer value"—the entire package of previous purchases. Others believe that customer revenue in a recent period alone is a better predictor of future

customer value. One academic study claims that a third metric, "customer lifetime value," is valuable for distinguishing the best customers.[23] Customer lifetime value itself can be measured in different ways.[24] Its basic components include the frequency of purchases, the amount of money the customer spends, the marketing resources allocated to the customer, and the likelihood that the customer will continue in the relationship.[25]

Financial establishments were among the first businesses to recognize the importance of evaluating the worth of customers. As database analysis became increasingly sophisticated in the 1980s and the 1990s, many executives at banks, at credit-card companies, and at brokerage firms saw real value in figuring out which of their customers were making them money and which were costing them money. One bank executive recalled that the surge in interest in databases became apparent around the beginning of the 1990s. He mentioned three factors external to banking as having helped drive the trend: the growth of sophisticated direct-marketing techniques, the founding of agencies that claimed expertise with database marketing (such as Ogilvy & Mather Direct), and the great increase in telemarketing. These developments, he said, had made banking leaders aware that they could use data more efficiently. "Sending [the same] letters to everyone just wouldn't cut it anymore," he added.[26]

Because banks have traditionally collected a lot of data on customers and their activities, the challenge is to analyze the information so that it helps to distinguish the kind and amount of value different patrons bring to the organization. To bring order to their understanding of customers, banks divide them into niches. Banks differ in how they do this. The vice president of customer-relations management at Virginia-based Riggs Bank, for example, noted that Riggs develops information on every customer from internal databases and from outside information sources. As important outside sources he mentioned credit rating companies and the database firm Choicepoint, which collects and sells wide-ranging information on millions of Americans. Nevertheless, he said, after analyzing all this material, the bank places its customers into only a few segments. "If you have more than five segments, you've got too [many]," he asserted. His reasoning was that too many segments made it hard for bank employees to understand their customers and know how to act toward them.[27] In contrast, executives at Pennsylvania-based Sovereign Bank said that buying personal data on their millions of customers is typically too expensive and that it is unnecessary. Instead, Sovereign relies on the 42 segments of the Claritas P$ycle database to link the knowledge it has of its customers to information that Claritas continually collects about people who are statistically like them. P$ycle, says Claritas, is "built from . . . an annual, proprietary national syndicated survey of more than 100,000 households"; it is "the largest database of household level consumer financial behavior."[28]

The aforementioned Riggs Bank official, who uses P$ycle secondarily to other proprietary databases, noted that its strength is in the data it presents about households that are statistically similar to the bank's patrons. The Sovereign Bank executives asserted that they use it for two purposes: acquiring new customers and selectively targeting existing customers. "The theory is that your new customers will highly resemble profiles of existing customers," he asserted. "So to market to get new ones you have to know what your old customers are like and [how] to keep your old ones."[29] P$ycle helps them figure out how to do that by statistically linking their customer to what Claritas knows about types—segments—of people it concludes are like them. When fed a bank's customer data, P$ycle software segments them "by evaluating the economic and demographic factors that have the greatest effect on their financial behavior."[30] That includes total household income, age of household head, home ownership, and "Claritas' proprietary measure of Income Producing Assets" (which it calls "a key predictor of real worth . . . for marketers that need to go beyond net worth and gross assets").[31]

The eight major groups into which P$ycle divides the population range from rich to virtual penury: Wealth Market, Upscale Retired, Upper Affluent, Lower Affluent, Mass Market, Midscale Retired, Lower Market, and Downscale Retired. Most of these groups are subdivided. Downscale Retired, for example, includes Downscale Sunbelt Security (people with rates of high home ownership but low amounts of discretionary cash) and Downscale Struggling Seniors (for whom day-to-day survival is an issue). The trick to making use of all these groups and segments, according to Claritas, is to link the data to the bank's "house file" to create "actionable" information. For example, the model finds that a household designated as Upper Affluent probably buys high-balance mutual funds, is more likely to have a home-improvement loan, and is likely to have a gold card with a revolving balance. A Lower Market household probably has a basic checking account and may have taken out a student loan. Inner City Strugglers, a segment of the Lower Market group, "tends to use basic financial services." Claritas adds that Lower Market households—who, it says, "can also be called the working poor"—are located in downscale urban areas, watch a lot of television, and listen to a lot of radio.[32]

David Kearsley, Sovereign Bank's Community Banking Manager, noted that his firm uses the kind of information just described to gauge the chances that a customer will have or need certain financial products. If Joe opens a checking account, he said, Joe is now part of the Sovereign database. At the end of every month, Sovereign's customer files are analyzed by Claritas, and Joe is placed in a group on the basis of the bank's basic information about him. The result is that the bank gets information about Joe's "product propensity"—his inclination to purchase certain financial products. "Let's say Joe is a P$ycle code 5 [a High-Asset Suburban Boomer]," said Kearsley. "Ninety percent chance of a mortgage. Uses internet banking a lot. High degree of likelihood of second mortgage." Knowing these propensities, bank personnel can look to see if Joe has those and related products with Sovereign. If not, the bank will reach out to Joe by postal mail, email, and telephone and in the bank to show him offers that can serve his life cycle's financial needs. In view of his high desirability, "the bank will credit-score him and prequalify him for a credit card and for equity." Everyone at the bank who interacts with him—on the phone, online, or at the bank—will see his code on their computer. The goal, said Kearsley, is to "build a solutions-oriented sales culture" that surround Joe at every touchpoint with bank materials customized for people like him.[33] People with P$ycle code 8 (Metro Elite Boomers) would get a different set of offers, as would those with P$ycle code 33—Young Urban Renters, who "are downscale but young . . . struggling to find their place . . . have low usage rates of many products but are heavy users of alternate delivery channels and loans."[34]

Bank executives score niches and then set up priorities as to what resources their banks will devote to them. A major issue among retail bankers is what to do with customers who do not deposit much money or buy bank products such as mutual funds and insurance but cost the bank money because they use bank services a lot. One obvious solution is to locate bank outlets outside neighborhoods with low-value clients. Even banks in moderate- and high-value areas, though, have clients who are low-value because of their socioeconomic position or because of their low spending with the bank despite their attractive niche.

Visits to tellers are of particular concern. Most banks, according to one consultant, see such a visit as "a high-cost customer service event."[35] In *American Banker* this consultant asserted that banks cannot allow value-destroying or low-profit customers to use their most expensive channels. He also noted that "too often we see value-destroying customers receiving top-level attention." The option is simple, he said, reflecting the general industry perspective on what he called "branch hogs": "Either they receive less attention or buy more products to increase their value to the bank."[36] Bank executives understand that this is easier said than done. They still recall that in the mid 1990s the First National Bank of Chicago was excoriated in the media and by public advocacy groups for charging low-value clients for using tellers instead of ATMs.[37] Although banks are not so obvious about this type of triage today, many have instituted routine charges for customers who visit tellers often. Other banks have different pricing schedules depending on whether people use kiosks or the web ("self-service channels"), visit, or telephone, and on how much value the customer brings to the bank.[38]

Citizens Bank, based in Rhode Island, offers five different "checking account products," whose benefits rise with customer value. The bank's website suggests that the benefits depend on where the customer lives. The bank demands a customer's ZIP code before providing comparative information on checking accounts. In the Philadelphia area, the lowest-tier "Basic Checking" requires a $10 minimum balance to open and has a monthly service charge of $3. For that, the bank allows ten withdrawals or checks and then charges a 60¢ "per debit fee" for each withdrawal transaction, including those at ATMs. This clearly is intended to keep low-value customers away from branch offices unless they are depositing money (for which there is no charge). Citizens Bank does allow "free online banking and bill payment" with Basic Checking. That, of course, assumes that a customer has an internet connection. With the top-of-the-line "Circle Gold Checking with High Interest," in contrast, online banking and bill payment is just one of many "freebies." The bank charges a $20 monthly maintenance fee for such an account; however, the fee is waived if, as expected, the customer keeps a monthly combined deposit and loan balance of $20,000. The fifteen other proudly displayed provisions include an exclusive toll-free number for Circle Gold Customer Service, higher interest rates on standard certificates of deposit, the bank's highest annual rate on checking balances, overdraft protection, free official bank checks, free Circle design checks on your first order and all additional reorders, free ATM transactions "at over 280,000 MAC STAR Maestro PLUS NYCE or Cirrus ATMs worldwide," discounts on home equity and instalment loans, free standard American Express Travelers Cheques, and free "investment consultation" through Citizens Investment Services Corporation.

Citizens Bank's focus on worldwide access and on investment instruments is indicative of what kind of customers it wants to cultivate. The message of differential value for customers of different value is quite explicit. In describing the second-tier checking program, the website says that it is marked by "the personal service and respect that comes with being a Citizens Bank customer."[39] The description of the third-tier program says "We believe in offering our customers more for their money, which is why we offer Circle Checking—a unique checking account designed to give special privileges to customers who do more of their banking with us."[40] The description of the top-of-the-line product goes further: "Our highest level of relationship banking, Circle Gold Checking combines value and service by offering you money-saving discounts, preferred savings and loan rates, and priority service. Think of it as our way of saying 'Thanks.' A lot."[41]

And banks have other ways to thank customers. Different levels of high-profit customers, a bank consultant stated, should elicit active "calling on the customer, with the goal of retention and cross-sell" of other bank products. The customer with low profit potential isn't worth telling about bank opportunities; this customer "merits only reactive service with a focus on cost minimization."[42]

Another way to thank high-value customers is through less time spent on hold when they telephone. Hoping to reward their best customers, banks often prioritize incoming phone calls on the basis of what one banker euphemistically called the "complexity of their needs." Entering or speaking an account number activates a priority rule tied to the individual client. Software is also used to match preferred clients to specially trained customer-service agents.[43] The goal is not just to make the callers feel good but also to use the opportunity to present opportunities tailored to their lifestyles.[44]

In addition, banks track customers' changing circumstances and try to reframe the offers they get. It is all quite systematic, and quite surreptitious. And it is done in response to alleged customer power. "We are entering a world of one-to-one marketing, where customers will no longer tolerate a wholesale approach to financial services." wrote the chief brand manager at Riggs Bank in 2004. "We cannot expect to recognize top-line revenue growth without exceeding their expectations," he added. There is a strong sense of cultivating loyalty—what one banker defined as a willingness to recommend the bank to others. There is also an implication of trust—that the bank will treat "customers individually" and fairly to help keep them. The actuality of the trust, however, is undercut by a lack of transparency with regard to the information on which the niche marketing is based.[45] The bank doesn't want the customer to know that the relationship is structured, and that offers are made, by means of arcane scoring of the customer on financial risk, potential value, profitability, and level of commitment to the bank based on information the customer doesn't even know the bank has.[46] The idea is to "gain guidance on how to build proper value equations, pricing strategies, and sales and service delivery." That approach, it turns out, is what determines "the appropriate style and depth of communication."[47]

::

Many high-end urban retailers have been working hard to match the banks' selectivity and their sophisticated database marketing. The reason for this is that such retailers feel squeezed by two powerful forces: While discount stores are making it impossible to compete by emphasizing price, the high cost of advertising on mainstream television and radio, in newspapers and magazines, and on billboards makes it difficult for high-end urban retailers to reach their best customers without spending far too much money. To a Bloomingdale's marketing executive, the implication is obvious: "We only have 32 stores spread around the country in expensive media markets. We live and die on training and knowing customers."[48]

Bloomingdale's is a textbook example of the Pareto "best customer" theorem at work. The top 20 percent of the chain's 20 million customers account for 73 percent of its business. Moreover, according to executives, these customers buy at Bloomingdale's more than 30 times a year.[49] To encourage them to feel wanted as well as to encourage them to increase the volume and value of their purchases, the chain commissioned a customer-relationship management system that the chain tellingly calls Klondike.

The search for more gold from the best customers begins with data mining. Bloomingdale's keeps a database on the transaction records of all its customers, but it uses Klondike to focus on the 15,000 most valuable ones. Klondike contains household information that the company has purchased about them. The information allows Bloomingdale's to divide the valuable 15,000 into niches that determine what promotional materials and what discounts they will get. Just as important, Bloomingdale's uses the data to create a customized relationship between the customers and sales associates over the phone and inside the store. To do that, Klondike makes the data about these people available to personnel in Bloomingdales' telephone call center and on its sales floors. By swiping a best customer's credit card at a point-of-sale terminal (a cash register), a salesperson can get an overview of that customer's shopping interests, "with the ability to drill down quickly," according to Richard Levey of *Direct* magazine. The idea is to "enable salespeople to custom-build merchandise suggestions. Aggregate spending information atop each customer's file allows the floor rep to make snap decisions about offering special services. If a consumer who buys thousands of dollars' worth of merchandise every year wants to return an item that hasn't been stocked in four years, a salesperson knows to accept it with a smile."[50] Levey depicts a system trying to take every touchpoint into account to extend buying time. The database creates customized messages about special events—for example, a "Girls' Night Out" promotion—that are sent to the point-of-sale terminal. The sales rep will read it when entering in the customer's purchase and so will be able to mention it. That information will also be custom-printed on the customer's receipt.

Klondike also tells sales associates which customers should be contacted by telephone about store promotions. Bloomingdale's designates some promotions as "storewide," some as "targeted," and some as "cosmetics." "For a storewide special, such as free gift wrapping, the associate with the greatest amount of interaction with the customer usually makes the call. If a promotion is contained within a single department—men's clothing, for instance—the rep in that area who has dealt with the customer is likely to [call]. And for cosmetics, a category that often engenders a relationship with a single salesperson, that individual is given exclusive access to the customer."[51]

Bloomingdales' head of marketing emphasized the importance of this sort of customer-relationship management for the chain. The retail consultant Karl Bjornson agreed and added that retail stores' use of customer information and segmentation to treat customers differently is becoming more advanced every day. He sees the activity taking hold across product categories, from apparel to consumables to and durable goods (including automobiles).[52] He and other experts acknowledge that it is difficult to specify the number or the nature of the segments that different retailers use. Every company has its own proprietary process, though often they are based on common database "engines."

According to database executives, whereas small firms divide their customers into only a few segments (and often only two—"men" and "women"), mid-size and large retailers often categorize customers by the channel they use (web, catalog, phone, or store), by how and where they were acquired as customers, by gender, by age, and by psychographics. With regard to psychographics (attitudes toward products and toward life), different retailers use different models. Income and geography are important to many retailers. So are ethnicity and race, which retailers feel have huge effects on buying decisions although they hesitate to say so for publication. Large retailers keep data for three to five years.

Bjornson said that department stores, in translating the data into ideas that marketers and sales associates can use, tend to categorize various groups by names that stand for the combinations of income, age, and position in life that analysts believe drive purchases of the products the retailer sells. In the apparel world, the names are invariably female because the prototypical consumer is a woman. "Elizabeth" may stand for one set of characteristics, "Sally" for another, and "Cindy" for another. In electronics retailing, the names are mostly male.

Sending particular types of customers' information about events and sales that match their buying profiles is called "pre-selling." The information might include coupons exclusive to the segment as well as promises of discounts at the point of purchase. When individual customers are subject to special treatment in a store, it is called "clienteling."

Pre-selling is more widespread than clienteling. It begins with frequent-shopper, loyalty, or credit cards, or some other way for the store to know the customer. Shop-Rite, a chain of supermarkets, encourages its customers to accept loyalty cards that allow them to get special deals; it then uses card swipes at checkout to track all their purchases. For privacy reasons, Shop-Rite doesn't purchase any data about its customers. It does, however, carefully sort through each shopper's buying habits. This continual "market-basket" analysis infers characteristics of the buyers and their households, with the aim of sorting them into niches. There are natural organic customers, families with young children, kosher customers, empty nesters and more. Though Shop-Rite doesn't share these categories with those who are slotted into them, it does have "clubs" which it encourages people to join—for example, there is one for parents with toddlers. Shop-Rite uses the data to pre-sell the right people with the right offers. It sends different mailings with different incentives to the different customer groups. It even sends special mailings and discounts to desirable customers in the database who haven't been in a Shop-Rite store for a while.[53] In addition, people who have different shopping patterns or belong to different segments may get different discount coupons for their next visit. Based on what the store knows, different products or different discounts may be promoted.

Separately, Shop-Rite uses "Catalina terminals" for coupon discrimination. Using revenue-sharing deals, Catalina Marketing arranges to place printers that issue coupons at the checkouts of stores throughout the United States. Manufacturers buy exclusive time for their product category on the Catalina network in four-week chunks. That makes its coupon dispenser a nationwide customized advertising medium. In "transactional advertising," a purchase of Tropicana orange juice may trigger the printing of a coupon for Minute Maid, that month's juice advertiser. In "historical customization," a Catalina coupon is based on a complex analysis of the consumer's purchases using the loyalty card over 104 weeks. Catalina does not know the customer's name; only an identification number and the purchase habits are recorded. The company's analytics group, however, determines statistically what kinds of buying patterns should be offered what kinds of coupons, and at how much of a discount.[54] Hypothetically, a person who bought Purina dog food two weeks ago but bought no dog food today might be given a coupon for Iams dog food. A shopper purchasing Iams today may receiving a coupon of higher or lower value than the other buyer's coupon, depending on the marketer's strategy toward its current customers. The supermarket, Catalina, and the advertisers expect that customers will save the coupons and use them on their next visit to the same Shop-Rite store.

Shop-Rite's approaches to pre-selling its segments and discounting on the basis of current and previous purchases are quite common for large chains. Consultants who talk about a step beyond that mention the CVS Pharmacy chain's ExtraCare loyalty card, which is carried by more than 30 million customers. According to a 2003 *Brandweek* article, CVS "knows a bit more about its customers than do most of its brethren." The article added that "the stores' familiarity goes beyond the clerk who sometimes greets customers by name after swiping their card at checkout."[55] Each sale becomes part of a complex database of customers' purchases that include prescription as well as nonprescription purchases and ages of children in their household. The database also knows whether ExtraCare members indicated on their card applications that they wanted to receive health information about women's health, diabetes, or general wellness. As a result of its ongoing data mining, CVS sends ExtraCare customers in various segments fliers that includes coupons with offers based on the recipient's prescription and non-prescription purchase history. People who buy lots of batteries from the chain might be rewarded with a "buy one, get one free" offer from Duracell. Those who inform CVS of their personal medical circumstances, or reveal them through prescriptions and other purchases, may receive catalogs of certain products—for example, CVS once sent out a 50-page mail-order catalog targeting caregivers and older consumers likely to need walkers, incontinence aids, and related items.[56]

Learning about individual customers' lifestyles in order to slot them into appropriate niches and then talk to them in particular ways is on the agenda of executives who champion the use of radio-frequency identification tags on individual products. RFID tags are miniature transmitters that contain electronic product code (EPC) data on an item, such as its universal number, its cost, its date of manufacture, and its date of shipping. Hand-held RFID readers transmit a low-power radio signal that causes any RFID tag in range to "power up" and exchange its EPC with the reader. The reader can then send the EPC to a computer database.

The Auto-ID Center, founded in 1999 at the Massachusetts Institute of Technology and working with three research universities and more than 50 global companies, focuses on "designing, building, testing and deploying a global infrastructure that will enable computers to instantly identify any object in the world."[57] So far, RFID tags have been used by Wal-Mart and other companies to track the movement of large pallets of inventory.

(Without opening the crates on a pallet, shippers can determine what is in them, where they came from, and when they entered transit.) Some retailing futurists, though, have pushed the idea of tagging individual consumer products so as to track their use. According to the magazine *Risk Management* in 2004,

The universal database for product codes that the Auto-ID Center has proposed and is actively working on would create a unique identifier for each product. When a consumer purchases any tagged product, that item's RFID code could potentially be associated with that consumer's credit card number. In fact, the use of unique ID numbers could lead to the creation of a global item registration system in which every product is identified and linked to its owner at the point of sale or transfer. For example, if the consumer purchased an article of clothing at a store, then returned to the store wearing the article of clothing, employees at the store could theoretically know the identity of the consumer. (Washable RFID tags that can be sewn into clothing are already available.)[58]

Risk Management further pointed that "small RFID readers could be embedded into carpets, floor mats, floor tiles, and doorways, allowing companies (or others) to continually monitor individuals wearing or possessing any thing with RFID tags. Some people predict a seamless network of millions of RFID receivers strategically placed around the globe in airports, seaports, highways, distribution centers, stores and even private homes, which would be constantly reading, processing and evaluating consumer purchases, behaviors, and even locations."

Experiments with RFID tags on individual items already have taken place. In what may have been the most prominent of these experiments, Wal-Mart teamed up with Procter & Gamble to put RFID tags in Max Factor Lipfinity lipstick containers. The idea was for P&G researchers to view customer's activities through video cameras placed near the shelves at the same time that they could note which specific lipstick containers had been picked up. The Chicago Sun-Times observed that the tags had a short read range—about half an inch—and that researchers could not track them or the people carrying them beyond the shelf. Nevertheless, the article noted, "manufacturers and retailers are looking at ultimately putting the tiny chips into everything from soda cans and cereal boxes to shoes, clothing and car tires."[59] The Sun-Times said it learned of the trial from a "disgruntled" Procter & Gamble executive. It quoted Kevin Ashton, executive director of the Auto-ID Center at MIT, as saying that "the idea that someone's privacy is at stake because there are a few RFID tags in a few lip-sticks in one store is silly."[60] Yet the reporters noted that the head of a privacy rights group, Consumers Against Supermarket Privacy Invasion And Numbering (CASPIAN), decried the practice. Moreover, the state of California moved to ban the use of RFID technology to track people as they shop or after they leave a store.[61]

Whether or not RFIDs become ubiquitous, retailers are looking for multiple ways to get good customers into their databases and to think of their store in ways that, analytics suggest, reflect each customer's lifestyle. Much of their work involves enticing people to voluntily enter into a spe-cial relationship with the store by signing up for a shopper card or a store-branded credit card in order to receive special treatment. Consumers who visit infrequently and don't have profiles that suggest they will ever become good customers might get fewer mailings, or none at all. In recent years, ever-more-specific analytics have also helped many retail outlets track customers who *lose* a store money by coming only to buy bargains or loss leaders. Best Buy found, for example, that approximately 100 million of the 500 million customers in its transaction-rich database were undesir-able because they used (perfectly legal) tricks to get discounts on top of dis-counts. To discourage such people from coming in, Best Buy instituted rules about returns and internet pricing that wouldn't make it worthwhile for bargain hunters to shop at Best Buy. The idea is even spreading to the supermarket, a domain of low margins and cutthroat competition that tra-ditionally has been afraid to cede any customers to rivals. In a world of Wal-Mart and analytics, however, even supermarkets have seen the utility of rewarding best customers with special service and prices and of discour-aging low-value customers from coming in. According to the CEO of the data-mining firm IRI, food and drug retailers have been compiling data from frequent-shopper cards for years but have been doing little with the data. That, he said, is begining to change quickly. IRI signed a deal with a major retailer to mine data from shoppers in order to help target market-ing toward the most profitable customers. He expected more supermarkets to do the same.[62]

An executive at another analytics firm argued that "retailers are less likely to overtly drive (unprofitable) consumers away." He added this, how-ever: ". . . if they're looking at discontinuing one of two items, and one is favored by their better shoppers and another by cherry pickers, when they make the choice, they de facto start that process."[63] In fact, some chains have taken steps to distance themselves from people who come in only for the discounts and spend relatively little. Dorothy Lane Markets received publicity in the grocery trade when it developed a "top customer-centered strategy" that discontinued the use of weekly fliers (which had attracted the bargain hunters) with the idea of using the savings to reward its best customers with special gifts in the hope they would increase their purchases. A columnist in *Progressive Grocer* noted that a small but growing number of chains are pursuing strategies that both invite "very good customers" and pushing away "cherry pickers." He opined that "creating a profile of their customers and then performing triage on the market to save their most valuable purchasers" is a wise competitive stance in a Wal-Mart world, where "competing on price is out of the question."[64]

To make its desired customers feel as if a store is treating them particularly special, many retailers turn to "clienteling." Karl Bjornson notes that the cultivation of good and best customers has become standard with department stores. Sales associates greet frequent buyers as they enter their departments, pour them coffee or even wine in special rooms, and even offer them unadvertised discounts at the point of purchase. As with Klondike, special treatment and discounts depend on the store's assessment of a customer's value. Even less elite merchandisers are getting into the act. Best Buy, for example, conducts training programs for employees on how to identify members of its target customer groups by their shopping behavior, and on different strategies for selling to various groups.

"Clienteling," of course, involves knowing when a "best customer" is coming into a store. Best Buy's attempt at doing it through training amounts to hit-or-miss stereotyping. Though salespeople in expensive areas of department stores might be able to do it regularly, supermarkets and consumer electronics stores don't know who comes in. In supermarkets, although Catalina Marketing claims a redemption rate for its coupons of 6–7 percent[65] (versus 1 percent for newspaper coupons), the idea of offering deals to best customers coming into the store on the basis of what the databases say is alluring to retailers. An obvious way to do this is to get customers to identify themselves electronically as they walk in. That is exactly what two large supermarket chains, Albertson's and Stop & Shop, are trying to do. In 2004, Albertson's provided all its stores in the Dallas area with hand-held terminals.[66] By swiping a frequent-shopper card, a customer unlocks a Scan & Shop terminal so it can be taken through the store. Each terminal has a small display that can send marketing information to the customer—for example, when a deli order is ready. When the customer wants to purchase an item, she simply scans it with the device. When she reaches the checkout, she slides the device next to a bar scan on the checkout. It alerts the terminal that the shopping visit is complete. No additional scanning is needed. The information scanned in the aisles is sent to the checkout. The end of the trip is just for paying. Custom messages and prices are delivered at the start of the walk through the supermarket; the device checks the customer's shopping history and displays discounts she can get through the store. Other deals may come up in the aisle, when the customer scans certain products. The basic idea is very much like that of Catalina terminal, but with the strong advantage of sending discounts at the start of and during the shopping experience rather than at the exit.[67]

Stop & Shop's experiment with shopping terminals is much smaller than Albertson's but more ambitious. It uses a wireless touch-screen IBM computer on the shopping carts of its test stores, of which there were twenty in 2004. Like the Albertson's terminal, this "Shopping Buddy" allows a person to swipe a loyalty card and to scan items as they are placed in the cart. Like the Albertson's model, it also allows for rapid self-checkout at the end of the shopping trip. It goes beyond the Albertson's terminal, though, by making it clear to the customer that she is enjoying a customized experience based at least in part on previous purchases. The device acts as a personalized shopping assistant that can display the shopper's buying history and favorites, a shopping list that can be created at home and emailed to the store, favorite items that are on sale, and the shopper's loyalty program points and reward level. The "Shopping Buddy" tracks the customer electronically as she moves through the store. As she approaches certain items in the aisle, the terminal presents personalized offers, including coupons, on the basis of how the supermarket has tagged her.

Albertson's employees in Dallas privately say that the Scan program has been moderately successful, older people having a harder time than younger ones understanding how to use it. Stop & Shop's public descriptions of its cart-based system were ebullient after the initial small test. "Everyone is using the unit," said one executive. "We see grandparents using them, parents with children, single men and women—once they use it for the first time, they continue using it."[68] Stop & Shop's president said: "Grocery shopping will never be the same once shoppers begin using the features of the new Shopping Buddy." He didn't mention the chain's desire to get people to identify themselves as they begin moving through the store. Instead, he emphasized that the goal is to "save customers time" and give them "new personalized services."[69]

Karl Bjornson added a two other justifications for getting customers to identify themselves upon entering a store: feeling valued and protecting their identity from theft. To Bjornson, the Albertson's and Stop & Shop devices are unwieldy because they require physical activation. Ultimately, Bjornson believes, retailers will persuade their very good customers to identify themselves with much less effort through biometrics—perhaps with a fingerprint, but more likely with an eye scanner. Some supermarkets are already collecting shoppers' fingerprints for use in check-cashing services. In the not-too-distant future, Bjornson posits, submitting to finger or eye biometrics entering a store will be the thing to do for customers who want to be protected and rewarded by the retailer as members of a community.

Biometric identification upon entering a store "can be packaged as a greater differentiator to protect my core customer," Bjornson asserted. "The best value proposition is the age old desire to be recognized, to have value as an individual, not a number." He contended that "as retailers are competing for a shrinking pie, they have to provide their customer with a value proposition. Price isn't the differentiator. . . . Price is what I pay for it. In today's world, it is increasingly difficult to ensure you're getting the best price. Consumers look for the true value—what do I do with the product once I get it? What is it doing for me?"

"Biometrics is becoming more and more common," Bjornson noted. "If I were convinced as a consumer that it would lead to greater opportunity for myself, I would be inclined to do it. Look at the internet. In today's world you are already subject to a lot of attacks, and it's going to get worse. Quite frankly, I would use the [biometric] technology so that I would know that someone [else] would not using my credit card."

Bjornson added that, in the spirit of what is already happening, retailers will not treat all customers who offer their biometric data as equals. The best customers in the best niches will get the best deals. In contrast, "people not in the right segments will be left behind. They will not have as rewarding an experience."[70]

7 :: Issues of Trust

In April 2005 *Advertising Age* announced a series of articles that, it said, would "explore how the business of marketing communications is being disrupted by forces such as digital technology, consumer empowerment and fragmentation." In the first article, titled "The Chaos Scenario," the columnist Bob Garfield confronted the question "What happens if the traditional marketing model collapses before a better alternative is established?"[1] Less than a week later, the *New York Times Magazine* published an 8,196-word piece by contributing writer Jon Gertner about challenges to traditional television ratings and disruptions of media business models. "I wouldn't predict that Nielsen is going out of business," Gertner quoted the head of research at Starcom Media Worldwide (a big media buying agency) as having said. "But they are at a crossroads. And it's almost as if their business model is evaporating overnight."[2]

It is no secret that marketing and media practitioners are having a painfully difficult time comprehending what is happening to their businesses. It is also no secret that a common goal emerging in this environment is to find out as much as they can about desirable audiences in order to attract and keep them. Many members of the public have interpreted this sort of information gathering as a bid to invade their privacy, and there has been resistance. Marketers see the resistance as a threat, particularly when it comes from sophisticated advocacy organizations. The last thing they want is outsiders igniting controversies about database marketing that might scare customers away and derail nascent corporate solutions to tough competitive problems.

Concerns are flowing from a variety of quarters, and executives are parrying them with public claims and behind-the-scenes lobbying. The claims have serious holes, but marketing and media practitioners are fortunate to have the social environment as an ally in keeping the flaws mostly hidden and the public stress levels controllable. It is an environment in which consumers' knowledge about retailers' power over information is low, government agencies focus mostly on scams and on narrow meanings of privacy, and advocacy groups do not get much coverage in the popular press for their opinions on database marketing.

Information accidents in the news—for example, the theft of database records and the rise of identity theft—may alarm consumers about the engine that is driving their "best customer" relationships. That may well threaten some database plans. Yet marketing and media players continue to cultivate this environment to their benefit, playing up useful aspects of database marketing and downplaying problematic ones.

Marketers and new-media developers are betting that the promise of better protection along with the warmth of friendship, service, and value in a harried, scary world of too many choices will override these concerns. They think that consumers, wanting to feel wanted and to get better value, will voluntarily join marketing and media lists. They believe that a predictable future will come to firms that create an image of trustworthy havens for desirable customers who raise their hands and provide personal information—even as those firms use the data to track and categorize customers without their knowing it, and to slot them into niches they may recognize vaguely or not at all.

::

Ignorance and tension in regard to databases need not be left to individuals to resolve. A number of organizations claim to be watchdogs or advocates. Various state and federal entities have drafted laws, pursued lawbreakers, or held hearings to highlight issues and to jawbone executives on various issues that have made it onto the public agenda. The Federal Trade Commission has claimed to be "the nation's consumer protection agency."[3]

Consumers visiting FTC.gov in mid 2005 would easily discern three related concerns: fraud, intrusions, and privacy. These categories accounted for the three largest items at the top of the home page: a tab on the left labeled "For Consumers," a large banner reading "National Do Not Call Registry, and a tab on the right labeled "Privacy Initiatives." The left link led to a notice that "in this section of our website, you'll find publications with advice on avoiding scams and rip-offs, as well as tips on other consumer topics."[4] The "National Do Not Call Registry" section explained how to enter a telephone number into an FTC database of the phone numbers of consumers who do not want to be bothered by telemarketing calls. The "Privacy Initiatives" link led to an invitation to read "about our efforts" to protect consumer privacy, including "what we've learned, and what you can do to protect the privacy of your personal information."[5]

The breadth of the FTC's mandate regarding database marketing is evident from its many activities having to do with consumers' personal information in the marketplace. These include monitoring the success of the Children's Online Privacy Protection Act, maintaining a database of identify-theft cases. and releasing a report on the privacy implications of radio-frequency identification. The FTC has also prosecuted companies for mishandling of personal information. In 2005, for example, it charged two mortgage companies with violating the agency's Gramm-Leach-Bliley Act's Safeguards Rule by not having reasonable protection for customers' sensitive personal and financial information.[6]

Notwithstanding the Federal Trade Commission's consumer-protection stance, it is a creature of the executive branch, which appoints the commissioners. Its website reflects a delicate line the staff must always walk between acting as a consumer advocate and allowing businesses to exploit information. The "education" areas of its website don't point out that privacy laws shield consumers less than they could. One matter not directly addressed is that the Gramm-Leach-Bliley Act permits bank holding companies to share some types of personal data among "affiliates" whether or not the customer gives permission. Nor does the FTC's description of the Gramm-Leach-Bliley Act underscore that when customers do get permission to decide whether affiliates should have access certain data, they must opt out (that is, request that the bank not use the data) rather than opt in (that is, affirmatively allow the bank to use the data).

The FTC's site reflects an arguable definition of the kinds of information privacy that deserve attention. The focus is on deterring the misdeeds of rogue businesses that illegally use or negligently release data that can be connected to a particular, personally identifiable individual. Non-personally-identifiable data rarely make it on the FTC's list of concerns. FTC.gov's "education" sections don't note that marketers increasingly use non-personally-identifiable data to mark, separate, and discriminate against people in ways that some might find socially problematic. Reports or public comments may highlight this development, but because it is not illegal the FTC doesn't highlight it.

The FTC's understanding of its mandate, then, frames privacy in ways that deflect attention from these and other criticisms of the U.S. government's current privacy policies. To learn about them, one must turn elsewhere. Available on the web for those who search are strident voices that portray government conspiracies with corporations to create a totally controlled society. For example, Alex Jones's "infowars" and "Prison Planet" websites include links to writings on government-mandated identification and attribute transaction-tracking systems to "the globalists . . . setting up the beast system so there is nowhere to hide."[7]

Jones and others with similar concerns have little patience with the belief that governmental or corporate policies on information privacy will ever come close to being acceptable. Worrying about some of the same broad issues are several advocacy organizations with much less of a conspiratorial mindset and more of a stake in mainstream political realities. Three organizations that work to critique marketing in this vein are Commercial Alert, CASPIAN, and Junkbusters. They place different degrees of emphasis on databases. CASPIAN focuses most on them. Its very name (standing for Consumers Against Supermarket Privacy Invasion And Numbering) refers to the security of customers. "We feel," writes founder Katherine Albrecht on the group's website, "that information about intimate details of our lives, such as the very food we put into our mouths, should not be stored in a computer database and subject to scrutiny."[8] CASPIAN's website vigorously champions the idea that frequent-shopper cards are ethically wrong and lead to price hikes by supermarkets. It excoriates the Kroger supermarket conglomerate for its shopper-card strategies, and it encourages a worldwide boycott against Gillette for putting radio-frequency identification tags on certain products to keep track of inventory and to prevent theft.

When I visited CASPIAN's site in mid 2005, much of the material hadn't been updated for nearly two years. Links didn't work, and the latest stories were weeks old. Commercial Alert's site, in contrast, is updated several times a month. Commercial Alert was founded in 1998 by Ralph Nader and Gary Ruskin. Its concerns are wide-ranging, and databases are not at the top of its priorities. According to its website in April 2005, the "mission is to keep the commercial culture within its proper sphere, and to prevent it from exploiting children and subverting the higher values of family, community, environmental integrity and democracy."[9] Database issues did not make the organization's home page of "Top Campaigns" on several days in which the website was checked during mid 2005. Junk food in schools, the role of teen magazines in adolescent suicides in the United Kingdom, and "annoying TVs on trains" were the centerpiece of discussion for April 21, for example.[10]

Commercial Alert did, however, discuss database concerns as part of its "culture" program link on the home page. It noted that "commercial list-brokers have targeted our nation's children" and urged visitors to fill out an email letter urging their congressional representatives to co-sponsor the Children's Listbroker Privacy Act, which, the site said, "would prohibit companies from selling the personal information of children below 16 years of age without parental consent."[11] Moreover, on the "about us" part of its site Commercial Alert took credit for two database "victories." Both were several years old. One was its successful campaign for the provision in the Elementary and Secondary Education Act that requires parental notification before a corporation can extract market research from a child in school. Another was its "campaign against commercialism in schools led to the demise of the ZapMe! Corp. and its use of computers to extract market research from unsuspecting school children."[12]

While Commercial Alert aims to oppose offending politicians and companies, Junkbusters offers to give consumers the power to go below marketers' radar screens. "Our mission," says the group's website, "is to enable you to get rid of any junk mail, telemarketing calls, junk faxes, junk pages, junk email, unwanted banner ads and any other solicitations that you don't want, while still allowing or even encouraging whatever you do want. We provide detailed information on how to stop any company from sending you stuff you don't want." The website explains U.S. laws governing junk-mail , and it provides "to tell organizations not to data about you."[13]

Junkbusters, CASPIAN, and Commercial Alert are unusual among advocacy organizations in worrying specifically about privacy in the context of marketing. In most privacy activism, marketing is just one area of a larger struggle. Computer Professionals for Social Responsibility, Privacy

International, the Electronic Frontier Foundation, and the Electronic Privacy Information Center (EPIC) have all converged on issues relating to individuals' rights over personal information that make database marketing issues relevant. But "relevant" does not mean "of the highest concern." For three or four years after airplane hijackers destroyed New York's World Trade Center, the U.S. government's turn toward collecting enormous amounts of information about people in a bid to find and track terrorists led these organizations to place primary emphasis on the implications of government surveillance. By late 2004, privacy experts turned again to concerns about corporate surveillance as a result of the growth of huge data brokers such as ChoicePoint and LexisNexis, the theft and misuse of personally identifiable information from them, and the increasing sophistication of biometrics and radio-frequency identification in retail settings. Although in May 2005 the websites of Computer Professionals for Social Responsibility, Privacy International, and the Electronic Frontier Foundation still were not highlighting database marketing, they were placing position statements about database marketing on their websites.

The Electronic Privacy Information Center (EPIC) went substantially further. The organization realized that, when it comes to regulating information privacy in the business arena, state legislation has often been tougher than federal approaches. With the March 2005 opening of a San Francisco office to focus on state privacy issues, the organization made clear that EPIC West would focus on information privacy.[14] The first two "top issues" listed on EPIC's home page were Choicepoint's loss of personally identifiable data to thieves and the prospect that Google might one day decide to sell the personally identifiable demographic and behavioral information its GMail service and its "online community" Orkut had collected. The third "top issue" was a "privacy regime" (suggested by George Washington University Law Professor Daniel Solove and EPIC West director Chris Hoofnagle) centered on the responsibilities of both government and business to Americans' privacy.

::

With all the information that is online, and with the aforementioned groups and others voicing concerns, one would think that the general press would cover database marketing. A rigorous examination of "the general press" would today have to include daily and weekly newspapers, television networks, local television stations, magazines, internet news sites, and other media outlets. What exactly should the target topics be, how many months or years back should the search go, and by what criteria should the material be analyzed?

One rough way to get a sense of how the press has discussed database marketing is by means of the "news" area of the Nexis database. It includes a panoply of popular and trade newspapers, magazines, scholarly journals, and even transcripts from the major television networks' news programs. To get a sense of the coverage of a database "scandal," consider the reverberation of the February 21, 2005 announcement by the data broker Choicepoint that the personally identifiable information it held on more than 140,000 residents of the United States had been inadvertently sold to identity thieves.[15] Type "Choicepoint" into Nexis for the dates February 21 through April 21, 2005 and you get the message "This search has been interrupted because it will return more than 1,000 documents." Change the time frame to only a month beginning February 21 and you still get more than 1,000 documents. Just the seven days from February 21 to February 28 yield 501 pieces. Looking through the articles during those seven days, though, reveals that the overwhelming number of them discussed the incident narrowly as identity theft. Many writers wondered why a company holding such sensitive data could be so gullible or careless. The mood was well characterized in the February 28 issue of *Information Week*: "The angry reactions to ChoicePoint's revelation that its database of personal consumer information had been compromised led politicians, in bandwagon fashion, to promise committee hearings and offer up improved legislation to enforce stricter privacy measures on companies dealing with consumer data." With a bit more distance, *USA Today* noted on April 11 that "data breaches at ChoicePoint, LexisNexis, the University of California and elsewhere, in which the personal records of thousands of Americans were pinched, underscore the brazen tactics of criminals marauding like gunslingers on a lawless internet, security experts say."[16]

The general press coverage of the Choicepoint incident, then, certainly rang the alarum on private database firms' vulnerability to theft of personal information and the possible need for the government to do something about it. But few of the articles in Nexis that mentioned Choicepoint during the first week of coverage stepped back to examine the firm's role in database marketing—for example, in niche creation and targeting. Only 75 (15 percent) of the 501 articles even used the word "marketing" or "marketers." Critical organizational voices that might have presented insights beyond the obvious privacy-theft angle were slighted. Only seven of the 501 articles mentioned the Electronic Privacy Information Center. Not one mentioned the Electronic Frontier Foundation, Privacy International, or Junkbusters.

Getting a broader sense of the extent to which critiques of database marketing made it into the press requires a longer time frame in Nexis and different search terms. Type in "Federal Trade Commission and database and (marketing or marketers)" for the two-year period ending April 26, 2005 and you get more than 1,000 documents. Further exploration of those documents (by splitting the search into months) reveals that a great proportion of the entries are Federal Trade Commission papers that few members of the public will ever see. A substantial percentage of the rest are from trade magazines with interests in federal information policies—magazines such as *Security Management* ("Rising Trend of Fraud, ID Theft" was the title of one article), *Information Week* ("Data in Peril"), and *American Banker* (in its "Regulatory Roundup," for example). Also present but in much smaller numbers are academic vehicles such as the *Journal of Marketing*.[17]

Examples from the popular press include informational articles such as the *Seattle Times* "consumer's guide," which advises: "Starting Dec. 1 [2004] you will be able to check your credit report annually for free. The new law requires the three major consumer-reporting agencies—Equifax, Experian and Trans Union—to provide a free credit report once a year to consumers who ask. Contact the Federal Trade Commission for details, 877-382-4357 (877-FTC-HELP); www.ftc.gov."[18] There also are pieces that combine alarming data by the FTC with suggestions about how to "act fast to prevent, limit damage from identity theft."[19] Other examples concern developments in which FTC regulations appear relevant, such as the refusal of some stores to accept returns by some people when the stores' databases show "excessive returns." A *Washington Post* article cites the FTC-administered Fair Credit Reporting Act in relation to consumer's right in this case but notes that "increasingly, companies are creating databases not envisioned by such regulations, and there is debate about which laws, if any, apply."[20] And there are stories of scandalous database schemes that are within the bounds of the law. A story in the *Chattanooga Times Free Press* titled "List Brokers Selling Children's Personal Information" exclaims "Commercial list brokers are selling the personal information of millions of children as young as age 2. And for now, it's legal."[21]

Although stories such as those mentioned above occasionally appeared in the news, mentions of the Federal Trade Commission and of database marketing spiked during certain periods as a result of revelations about criminals who stole marketing databases and about database-marketing firms that failed to keep data secure. Stories in August and September 2004, for example, revealed that the Federal Bureau of Investigation, the Federal Trade Commission, and the Postal Inspection Service had announced the arrests or convictions of more than 150 people, including a Ukrainian man who allegedly used internet chat rooms and his own website to buy and sell stolen credit card data, in a nationwide crackdown on internet fraud.[22] Three high-profile cases appeared in 2005. The first, already noted, involved the purchase by thieves of personally identifiable information held by the Choicepoint database company. It was followed by infiltration of data about 310,000 people in a LexisNexis database subsidiary, and still another in which data about millions of credit card, debit card, and check transactions were stolen from the computer system of DSW Shoe Warehouse over several months.[23] While reports of the Ukrainian and DSW incidents blandly cited FTC regulations and advice, some articles and editorials about LexisNexis, like the earlier ones cited about Choicepoint, questioned the propriety of allowing companies to handle so much personal data without coming under the FTC-administered Fair Credit Reporting Act.

Thus, while it certainly was possible for close readers of the popular press from April 2003 through April 2005 to learn how new styles of database marketing were affecting consumers' lives and what that might mean beyond identity theft, that wasn't a major theme. Although the general press cast up tales of miscreant database firms and frightening case studies of identity theft, it gave its audience little explanation of the rules about who is allowed to control information about them in the online or the offline marketplace, or about the extent to which laws prohibit price discrimination based on behavioral targeting. The phrase "behavioral targeting," in fact, hardly appeared in the general American press during those two years, to judge from Nexis. Although the phrase yields 332 entries between April 26, 2003 and April 26, 2005, almost all were in marketing trade magazines. Several mentions turned up in non-U.S. publications. Only four mentions appeared in American dailies.

"Price discrimination" did not fare much better. Although it did occur in 943 Nexis documents, the great majority of those were in trade magazines or government journals. The remainder tended to show up in foreign English-language newspapers; very few appeared in general U.S. newspapers or magazines. When the phrase "supermarket or retailing" was added to "price discrimination," the number of mentions dropped to 156, and only five were in general U.S. newspapers or magazines. Three of those were actually the same op-ed column, published in three different papers, and that column was strongly in favor of price discrimination.

In view of this rough overview of press coverage, it is not surprising that many Americans were in the dark about major facts of life affecting them in the emerging marketplace. Two national telephone surveys conducted by the Annenberg Public Policy Center confirm that characterization.[24] One of these surveys, carried out in early 2003, centered on the internet. We asked a randomly selected sample of 1,200 adults whose homes had internet connections questions that explored their understanding of how firms handled information about them online. The other survey, carried out early 2005, dealt with merchants' information activities offline and online. We asked 1,500 randomly selected Americans who had used the internet in the past 30 days (at home or elsewhere) how much Americans knew about who was allowed to control information about them in the online and the offline marketplace. And we asked what they knew and felt about behavioral targeting and price discrimination offline and online.

The 2003 survey provided strong evidence that the overwhelming majority of American adults who use the internet at home have no clue about data flows—the invisible techniques whereby online organizations extract, manipulate, append, profile and share information about them. Fifty-nine percent of adults who use the internet at home agreed with the statement "When I go to a website it collects data about me even if I don't register." This basic knowledge of cookies did not generalize to deeper understanding, though. For example, 57 percent stated incorrectly that when a website has a privacy policy it will not share their personal information with other websites or companies. The ignorance about privacy policies is, however, only the tip of the iceberg of confusion about what goes with personal information behind the computer screen. The reactions of most online-at-home adults to a common way websites handle visitors' information indicate that they do not understand the collection, interrelation, and use of identifiable and anonymous data.

We presented our interviewees with a supposed change in the information policy of a website that they had previously said they "like most or visit regularly from home." The goal was to gauge the acceptability of a common version of the way sites track, extract, and share information to make money from advertising. We read the version to five web experts from academia and from business, government, and social advocacy groups who agreed that what we would be presenting was a common example of a site's approach to information. Accordingly, we integrated the hypothetical scenario into a questionnaire that asked people what they would do if a favorite website were to announce that henceforth they could access it for free only in exchange for allowing it to use personal information about them in order to make money from advertisers. We said: "It will learn about you by getting your name and main email address, by buying personal information about you, and by tracking what you look at on the site. The site will not directly tell advertisers most of the information it learns, though it may tell advertisers your email address. It will send ads to you for its advertisers based on the information it learns." The responses were straightforward. When presented with a rather common version of the way sites track, extract, and share information to make money from advertising, 85 percent of adults who go online at home did not agree to accept it, even from a valued site. When offered a choice between getting content from a valued site with such a policy and paying for the site and not having it collect information, 54 percent of adults who go online at home said that they would rather leave the web for that content than do either.

The widespread rejection of what is actually a common version of the way sites track, extract, and share information to make money from advertising suggests that adults who go online at home overwhelmingly do not understand the flow, manipulation, and exchange of their data while they are online and afterward. Other findings indicate that a substantial subset of the people who refused to barter their information is especially ignorant about information activities on the web. Among the 85 percent who did not accept the marketing deal, 53 percent had earlier said they gave or would be "very" or "somewhat" likely to give the valued site their real name and email address. Yet those bits of information are what a site needs to begin creating a stream of data about them—the very flow (personally identifiable or not) that they refused to allow in response to the scenario. Moreover, 63 percent of the people who said they had given up such data had also agreed that the mere presence of a website privacy policy means that it won't share data with other firms. Bringing these two results together suggests that least one of every three of our respondents who refused to barter their information either do not understand or do not think through basic data-collection activities on the internet.

The 2005 survey showed that this ignorance of the particulars of information use carries over to an understanding of price discrimination and behavioral targeting. The sample was different. Whereas in 2003 we surveyed individuals who said they use the internet at home, in 2005 we interviewed people who said that they had used the internet during the past 30 days. It turns out that 91 percent of them had home internet connections, and almost exactly the same percentages described themselves as beginners, intermediates, and experts when it came to using the internet. Still, the 2005 sample was a broader population and so may have been even less aware of goings on behind the screen than the 2003 sample. The questions we asked, however, went beyond the internet to the world of non-virtual supermarkets and the banks.

At the core of that nationally representative phone survey was a 17-statement true-false test about laws and practices of price discrimination and behavioral targeting and about where people can turn for help if their marketplace information is used illegally. On average, the respondents were correct on only seven of the statements. Most did not know who is allowed to control information about them that can lead to price discrimination. Most were also incorrect in believing that the law protects them from secret forms of price discrimination offline and online. Beyond factual misunderstandings, the survey revealed that internet-using adults overwhelmingly object to most forms of behavioral targeting and all forms of price discrimination as ethically wrong.

Some of what people didn't know can quite specifically lead to economic loss. Sixty-eight percent of American adults who had used the internet in the past month believed incorrectly that "a site such as Expedia or Orbitz that compares prices on different airlines must include the lowest airline prices." Forty-nine percent could not detect "phishing"—the illegal activity in which crooks posing as banks send emails to consumers that ask them to click on a link wanting them to verify their account. Sixty-six percent could not correctly name one of the three American credit reporting agencies (Equifax, Experian, TransUnion) that can keep them aware of their creditworthiness and whether someone is stealing their identity.

Incorrect answers to other statements indicate that consumers are also vulnerable to subtle forms of exploitation across a wide variety of online and offline locations. Sixty-four percent of the adults who had used the internet recently did not know that it is legal for "an online store to charge different people different prices at the same time of day." Seventy-one percent didn't know it is legal for an *offline* store to do that. Seventy-two percent didn't know that charities are allowed to sell their names to other charities without their permission. Sixty-four percent didn't know that a supermarket is allowed to sell other companies information about what they bought. Seventy-five percent didn't know the correct response ("False") to the statement "When a website has a privacy policy, it means the site will not share my information with other websites and companies."

Of all characteristics in people's backgrounds, having more years of education was the best predictor of understanding basic realities about power to control information about them and the prices they pay in the online or offline marketplace. Yet even having more general schooling doesn't necessarily mean really knowing this world well. People whose formal education had ended with a high school diploma knew the correct answers to an average of 6.1 items out of 17. People with a college degree did better (8.1 items), but that is only 48 percent. Even people with graduate school or more averaged only 8.9 correct (52 percent).

Despite the overall lack of knowledge, the people we interviewed were aware that companies could follow their behavior. Eighty percent knew that marketers "have the ability" to track them across the web. When presented with scenarios describing different types of behavioral targeting, 84 percent said that they believed that some websites analyze what people are reading, change the ads on the basis of that reading, and buy personal information about the readers from database companies. Eighty-nine percent of those who said that their supermarket offered a frequent-shopper card had accepted the offer—and, in order to get the card, had given the store information about themselves.

Our respondents did admit feeling vulnerable in a retail environment in which they knew they could be tracked. Only 17 percent agreed with the statement "What companies know about me won't hurt me" (81 percent disagreed). Seventy percent disagreed with the statement "Privacy policies are easy to understand," and 79 percent agreed with the statement "I am nervous about websites having information about me." All types of price discrimination drew strong objection. When presented with various concatenations of price discrimination, between 64 percent and 91 percent of respondents registered aversion to the activity. Interestingly, a smaller percentage (64 percent) disagreed with discount coupons as mechanisms for price discrimination compared to simply asking for less money (76 percent). The largest percentages rejected the prospect that different people would pay different prices for the same products in the same hour. Eighty-seven percent disagreed with that for an "online store," and 91 percent for a supermarket.

Most people also disagreed with behavioral targeting, but certain kinds of behavioral targeting yielded far more disagreement than others. Statements about keeping records on people's buying habits in order to decide what prices to charge them received far more negative responses than those that involved showing people different products on the basis of database information. In turn, buying data about people and showing them different products as a result was less acceptable than showing people ads online on the basis of what they were reading at the time.

In both the 2003 and 2005 studies, the overall sensibility that comes through is this: People are wary of firms' collecting information about them without their knowing it, and they want openness in their dealings with companies. Ninety-four percent of the 2003 sample agreed with the statement "I should have a legal right to know everything that a website knows about me." Eighty-six percent agreed that laws forcing website privacy policies to have a standard format would be effective in helping people to protect their information. Eighty-four percent of the 2005 sample agreed with the statement "Websites should be required to let customers know if they charge different people different prices for the same products during the same hour."

The respondents in both populations didn't hold out hope, however, that either business or government would take up their cause. Whereas in

2005 we found that 81 percent disagreed with the statement "What companies know about me won't hurt me," in 2003 we explored this question specifically with regard to the online world. We found that only 18 percent of those who accessed the web at home said they trusted their banks and credit card companies to protect personal information online while not disclosing personal information without their permission. Moreover, only 13 percent said they trusted "the government" to act that way. Unfortunately, distrust of the government's role carried over to the offline world: In 2005, only 35 percent of internet-using American adults said they "trust the U.S. government to protect consumers from marketers who misuse their information."

::

The picture of consumers that these and similar studies draw is one of nervousness and confusion—an openness to exploitation by merchants who decide what products to show and possibly what prices to offer informed by sophisticated storehouses of behavioral, demographic, and psychographic data. In the furiously competitive American retail marketplace, though, this view is not the one projected. Instead, executives portray consumers who are concerned about privacy as a powerful force. "If privacy blows up, that's the kiss of death for a retailer," asserted Dan Hopping, senior consulting manager for IBM's Retail Stores Solutions division.[25] His comment reflects a belief that worries about consumer privacy must influence how mainstream retailers conduct business. IBM has been a leader in thinking this through. Its Privacy Research Institute in Zürich defines privacy as "the right of individuals to determine when personal information can be collected and how it should be used based on individual consent. Unlike security, which revolves around the authorization of users, privacy addresses data management issues related to users who have already been given access to the system. Corporations need to handle this private information in compliance with privacy regulations as well as business requirements."[26] Building on this perspective, the Privacy Research Institute has suggested stringent "identity mixing" ("personal data is best protected by not being revealed at all") and a more tolerant yet cautious model of data mining that would enable firms to extract statistical data while safeguarding personal elements.[27]

In everyday business, IBM's privacy ethic seems to translate into helping companies make sure that customers "opt" to be included in the firm's database and that customers are correct in believing that personal information will not be viewed without a customer's permission. With respect to inducing customers to allow merchants to collect their information, Hopping advises that "the retailer's going to have to give them something." He tells IBM's clients that the inducement can be a discount but often is something else—special parking, services, or technologies like the Shopping Buddy that seem to make a supermarket visit easier. When it comes to the privacy of customer data, IBM focuses on conservative database management. It advocates the grouping of customers into categories that allow retailers to tailor (and even personalize) offers to members of a class but not to individuals. Although this does make the data private in the sense in which the word is typically understood (and most firms wouldn't even go that far toward losing the ability to see the individual), IBM's technique still allows for great discrimination among consumers through the sophisticated creation of "classes" or niches. To the extent that highly specific niches of consumers can still be identified (as IBM assures they can), even marketers who adhere to IBM's privacy ethic can exploit data in an advanced manner for the customization of advertising, information, and news. More troubling, IBM's valorization of privacy as leading to customer loyalty through trust takes place at the same time as activities aimed at the quiet undermining of trust. One thing IBM researchers are investigating with vigor—and something they surely would not advocate firms' telling consumers about—is how customer data might be used to figure out which individuals are not worth the trouble of wooing. In its Israel research lab, IBM is designing advanced statistical and machine-learning models that will differentiate customers according to their future value on the basis of a relatively small number of variables. According to *Technology Review*, the beauty of the model lies in its "domain adjustable" nature: "It could be used in banking to determine whether to issue a loan or a credit card. Or it could be employed by retailers to target promotions to potential best customers and give priority to those customers during times of peak demand."[28]

In trade-magazine quotes and at conferences, marketing executives seem to justify using consumers' personal information against them by depicting them not as confused and nervous but as knowing and aggressive, with ever-sharper tools that make *retailers* nervous. In December 2004, *Adweek* quoted Yahoo's retail category development officer Michael Schornstein as having said: "Consumers have become very adept at cross shopping—and that's when they use the internet to become educated on products, features, retailers, price."[29] A month later, the *Adweek* columnist Bob Greenberg was even blunter in describing customers' power: "Customers are making decisions in new ways today. They are players, taking control of how they engage with brands, wherever and whenever it is convenient."[30] These oddly contrasting pictures of consumer power are not necessarily mutually exclusive. It may well be that growing segments of the American population are finding websites and other vehicles that help them shop for good deals. It is also quite possible that these people have far less knowledge about the rules and emerging practices in the offline and online marketplaces than most merchants think. In any event, retailers seem to be operating as if they are dealing with savvy shoppers. That gives them the justification and the initiative to find ways to gain advantage over consumers who try to use digital media to gain marketplace power. The idea was captured especially well in this pronouncement about online product searching in the British trade magazine *Marketing Week*:

. . . search does signal a fundamental flip in the marketing environment. We are moving from an environment where structures and processes are controlled and driven by sellers and their search for customers, to one where the key mechanisms and content will be driven by buyers and their search for value. Because search pre-empts and informs every other shopping process, the more consumers take on the search habit, the more traditional go-to-market processes will need to fit these habits. Whether you work in advertising, direct marketing, advertising-funded media, retailing, or for any company that relies on these mechanisms to go to market, the secrets of success are set to change. Perhaps not this year. But very soon.[31]

Concern about consumers' new access to knowledge is high among many marketers. Nielsen/Net Ratings spread the news that one-third of all American internet users visited a comparison-shopping site in October 2004.[32] Responding to this sense of a swelling wave, trade press articles routinely profile high-traffic comparison-shopping websites such as Yahoo Shopping, Bizrate, NexTag, and Froogle as tools that can harm merchants or help them, depending on how savvy the merchants are. The dread comes from the sites' potential to introduce consumers to new sellers and increase competition. Comparison-shopping engines allow shoppers to research details of specific products, different retailers' prices for the products, and quality ratings of the products and retailers. If a consumer finds a desired item at a price he is willing to pay, the shopping site will direct him to the merchant's site to make the purchase. Or, armed with the data, the consumer may decide that buying the item in a physical store is the best bet.

In 2005, in *Chain Store Age*, a senior research analyst for the Hitwise internet research consultancy asserted that growth had "leveled the retail playing field and permanently altered how consumers shop, both online and offline."[33] "Comparison shopping," she continued, "is a natural behavior" for someone searching items on the internet, "given that a shopper can visit dozens of stores in a matter minutes." Hitwise data, she said, showed that about half of visitors to the "average shopping site" visited another shopping site first, and half of those would go to another site afterward. Moreover, the share of internet visits to the top ten comparison-shopping sites grew by 26 percent between December 2004 and December 2005. Part of the reason was the broadening of the types of products sold through the sites. Whereas at first they mostly listed high-priced electronic gizmos, most sites now show many shopping categories, from clothes to beauty products to computers.[34]

The tone of trade press articles about this development tends to be upbeat. The message is that retailers can turn comparison-shopping sites into benefits for themselves. "Comparison shopping engines can help you boost sales and acquire customers," said a front-page article in the January 2005 issue of *Catalog Age*. But the writer quickly added a caveat: "—provided your company's product is a top contender among the search results." Then came a discussion of what that means. The basics are that only Froogle (a subsidiary of Google) and Shopzilla offer free results. However, both charge for preferred placement—Froogle on the side, Shopzilla on the top of search results. The costs of participating in paid-inclusion comparison sites such as Bizrate, NexTag, and Shopping.com varies by product category. The typical approach is to charge the merchant from 5 to 10 cents per click to in low-margin categories such as books and toys, and $1 or more in high-margin categories such as ink cartridges.

Different sites use different classifications—typically, product, price, and shipping costs—to sort search results. Major shopping engines accept continuous data streams about products from the merchants, so files containing updated product information, pricing, shipping costs, and other details go to the sites automatically. Retailers should "evaluate the data often, and don't be afraid to 'slice and dice' the information to take out underperforming products."[35] Retailers also should know that consumer interest increases "with each additional bit of information you can provide them."[36] That may entail including a logo with the listing to stand out and apprising consumers of the total costs (including shipping and tax) so they will not be afraid to go a retail site's checkout. Consultants caution that these are not mere information-technology issues. Rather, says one, "by fine-tuning the feeds with better-written copy and an understanding of how the comparison engine categorizes products . . . you can increase a company's visibility on a comparison site."[37] And, according to the mantra, visibility often translates to cash.

One study found that the top slot on a shopping search engine can generate 40 percent more leads than the second slot.[38] Getting that top slot—by bidding high for a paid rank or working to compete in the unpaid area—can be a daunting goal. All seem to agree that the reason for pursuing it is not sheer customer numbers. The great majority of consumers still use regular non-comparison search engines such as Google to check out goods.[39] But the belief among retailers is that comparison-shopping sites bring shoppers who are ready to buy—"qualified shoppers," in the language of the trade—to the merchant's online location. Stores hope that the next time they buy similar products they will return there instead of going to the comparison site. Knowing that low pricing is a particularly strong way to get noticed on the comparison sites, some merchants charge less for products they post on comparison sites than they charge on their regular sites. "Overall," a pricing consultant notes, "you see a lot discounting on the web, less in stores, and least in catalogs."[40]

::

Product-comparison sites work both for and against big and small manufacturers and retailers. Help for the little maker or seller comes through an ability to see a wide range of models and prices that can be arrayed in multiple ways. Search on Bizrate.com for digital cameras, for example, and it is possible to compare the features of over 45 brands that make up hundreds of camera models sold by dozens of retailers. Many of the brands

(Largan, for example) wouldn't make it into most stores, and so the ability of the manufacturer and the stores that sell it to reach out to consumers shows how the terrain has changed.

There is, however, no denying that Bizrate.com and the other large comparison shopping sites privilege the big makers and sellers. Go to Bizrate's Digital Cameras main page and seven major brands stand out in the sidebar listing that links to specific information on Canon, Nikon, Sony, Kodak, Olympus, Panasonic, and Konica Minolta cameras. The sidebar names change, and their order is not alphabetical; the list seems to result from the firms' payment for the placement. To get to the other fifty-plus brands, one must click on the word "more." To further emphasize the power of top brands, cameras from the major brands take up greater part of Bizrate's Digital Camera main page. They descend in price, but Bizrate grants them all legitimacy by notes such as "found at 40 stores" and "33 reviews" with numerous stars denoting quality. Click on "more" and then on the word "Largan," and you will find one store that sells it, no product rating, and the dubious encouragement "Be the first to review this product!"

The presentation of retailers allows for a greater leveling of the playing field than the listing of products. Click on the name Nikon D70s, for example, and you get a list of 39 stores with "prices ranging from $748 to $1,450." The page initially presents its "featured stores" at the top of the list, but all stores are listed by—and can be sorted by—"five quality categories": "would shop again," "on time delivery," "customer support," "product met expectations," and "customer certified" (which means it has met certain minimum satisfaction criteria). Rankings are denoted by happy, neutral, or sad faces, and there are explanations, praise, and rants from alleged customers. Type your ZIP code into a box and you get the total price including tax and shipping. If you know you want a Nikon D70s, and you know what accessories you want with it, you may get a pretty good deal (relative to what you can get at your local camera store) from a seller that seems to have been vetted by Bizrate and other consumers. At the very least, it gives you a basis for comparison.

But the major retailers work the system to their benefit. Like the major manufacturers, or in league with them, the retailers can parade their personalities—their names and logos—by bidding enough money to stay atop paid rankings relevant to their products on comparison-shopping sites.

They can also spend the many dollars required to analyze competitors' prices so as to adjust their own on the comparison engines to draw customers to their sites. One method of adjustment that some stores use for cameras is to offer "packages" that seem optimal for people who link to their sites through particular comparative-shopping websites. The stores present different packages (often with more varied prices) for people who come to them directly, without first visiting the comparison sites.

The utility of this scenario for the store emerges if while at Bizrate you decide to purchase a Nikon D70s from Wolf Camera on December 24, 2005 at 5:20 P.M. Eastern U.S. time. The store is customer certified and has stellar ratings and more than 2,500 reviews. It lists the camera for $859.99 without shipping. Under "Notes From This Store," Bizrate informs you that the store sells "2 packages." Click on that link, and you are taken to another area of the Bizrate site that contains Wolf Camera links for the D70s, "D70s Digital SLR Body Only [without the lens]" and "D70s Bonus Outfit— Buy a Bundle-Save a Bundle." The camera body is the one for $859.99; the outfit costs $1,319.00. Both come with a "FREE Epson Printer Offer." Bizrate now gives you the possibility to compare outfit prices with other vendors. Let's assume that you decide to stay with Wolf to learn more about the deals. You click on the outfit link, and it takes you to the store website. You learn that includes a D70s body with an 18-70-mm Nikon DX lens, a Tamrac Digital Zoom 4 Gadget Bag, a 512-MB CompactFlash memory card, a replacement battery for the camera, and a one-year extended warranty that includes damage coverage. You also learn if you on a section titled "Incredible Offer" that the free Epson printer is not automatic. To get it, you must buy one of a few Epson printers that cost $150 when you purchase the camera, then send the proof-of-purchase to Nikon. The photo company will send a rebate check of $150.

You can try to find these particular deals at other Bizrate vendors to see what they charge; if you are satisfied with Wolf Camera's prices, you can purchase one of the packages from Wolf through Bizrate. But go to Wolf Camera through Google or by typing in its web address (www.wolfcam era.com) and you will get a different set of deals to consider at the Nikon D70s page. In addition to the "body only" offer and the "bonus outfit" with the Epson printer offer, the store presents two other proposals. For $1,199.99 it offers a D70s with the 18-70-mm Nikor DX lens and the "free printer offer" but without the bag, memory card, replacement card or the extended warranty. For $1,349.98 the store site offers the 18-70 lens bundled with an Epson Picturemate Express Photo printer and the added possibility of getting a rebate on another Epson printer.

Which of those deals is best, and how do they compare to similar offers on Bizrate? The answer would depend on individual needs and still more comparative investigation of other sites. Clearly, though, Wolf has decided to list only some of its Nikon D70s packages for its Bizrate customers without telling them. It's hard to understand the competitive strategy that led to not also offering the $1,199.99 basic kit, but the answer may have to do with Wolf Camera's profit requirements after sharing the proceeds with Bizrate as well as with an understanding of the people who visit Bizrate compared to their own sites. In view of increasingly sophisticated analysis of consumer activities on retail sites, it is not a huge step from presenting selective offers to presenting customers with different prices depending on how customers got to the site, what they have paid in the past, and what the store knows about them.

As I noted in chapter 6, banks, supermarkets, and other retail outlets already engage in price customization based on what they know about individual customers. Merchants consider the online environment particularly ripe for such "dynamic pricing"—that is, for price discrimination driven by behavioral targeting. In the *Harvard Business Review*, associates from McKinsey & Company chide online companies that they are missing out on a "big opportunity" if they are not tracking customers' behavior and adjusting prices accordingly.[41] Consultants urge retailers to tread carefully, though, so as not to alienate customers.[42] The most public revelation of price discrimination online centered on customer anger toward Amazon.com in September 2000 when it offered the same DVDs to different customers at a discount of 30, 35, or 40 percent off the manufacturer's suggested retail price. Amazon insisted that its discounts were part of a random "price test" and were not based on customer profiling. After weeks of customer criticism, the firm offered to refund the difference to buyers who had paid the higher prices.[43]

Though website executives are wary of discussing the subject, it seems clear the practice continues. Consumer Union's Webwatch project found many bewildering and seemingly idiosyncratic price differences, sometimes quite large, in its investigation of airline offers on travel sites.[44] When asked whether travel websites vary prices based on what they know about customers' previous activities, one industry executive told Webwatch advisor and University of Utah professor Rob Mayer "I won't say it doesn't happen."[45]

Another reason to believe that shopping sites are not consumers' counterbalance to retailer power is that the sheer complexity of decision making may make at least some people shy of using them, or of going beyond the most well-known manufacturers and retailers. Retailers understand this. A strong refrain in trade magazines' articles about comparison sites is that the lowest price is not necessarily the key determinant of sales. *Catalog Age* assured its readers in January 2005 that "even if you don't offer the lowest price on a comparison engine, most experts recommend getting in on the game." The U.K.'s *Marketing Week* reported in late 2004 on a Cambridge University study that "only one in eight online shoppers is solely motivated by price"—at least when it comes to the electronic and computer appliance items it studied on the major U.K. site Kelkoo.[46] More broadly, the Jupiter Research consultancy advised around the same time that about 70 percent of click-throughs on comparison sites "are not to the merchant with the lowest price."[47]

Kelkoo.com's U.K. managing director added: "A lot of people thought the internet could never be brand-driven, but this research proves that the most important factor is still a strong brand."[48] As if elaborating on the meaning of brand, a Jupiter analyst said: "Yes, people care about price, but it's not necessarily all about price. It's about the customer getting the best deal and getting the best product at the right price from the right retailer at the right time."[49]

Reputable merchants worry that retailers who exploit consumers through comparison shopping sites will scare consumers away from web shopping. At the same time, large retailers benefit from these stories when consumers draw the lesson that name brands sometimes trump low prices. An angry blogger's revenge on PriceRitePhoto.com is as an example of the speed through which web users can devalue a merchant's reputation while it also provides a cautionary tale about the importance of trusting an online seller. As "Thomas Hawk"—a technology and photography blogger based in San Francisco—tells it, he tried to buy a Canon EOS 5D in late November 2005, when he saw on Yahoo Shopping and on PriceGrabber that PriceRitePhoto.com was offering it for an unusually low price. He filled out PriceRitePhoto's order form on the web but was phoned by someone from that company suggesting that he order several camera accessories. When he said he didn't want to do that, PriceRitePhoto didn't process his order, which led to heated exchanges and threats on both sides. Hawk shared his experience on his blog and he posted a link to the story of a community-driven news site, Digg.com. Word traveled quickly, causing "thousands of people" to jam the website, send it viruses, and make prank phone calls to the firm.[50] A story in *OnlineMediaDaily*, a marketing trade magazine, drew the following conclusion: "PriceRitePhoto.com, a Brooklyn, N.Y.-based camera e-retailer, recently found that blogosphere justice can be swift, but is rarely merciful. Thursday, around 48 hours after 'Thomas Hawk' . . . posted a nightmare tale of hard sells, threats of legal action, endless delays, and runarounds, PriceRitePhoto.com has found its Web site in shambles, and its listings removed from prominent shopping aggregators like PriceGrabber.com and Yahoo Shopping."[51]

A few days after extolling the self-correcting nature of web retailing, *OnlineMediaDaily* went back to the Hawk experience to explore the danger that price comparison sites posed to customers because of fraudulent positive feedback merchants may create about themselves that make them look good on those sites. "One of the things that troubles me the most about this situation is that I found this retailer through Yahoo Shopping and they were perceived to have positive feedback," *OnlineMediaDaily* quoted Hawk from his blog. "Is the feedback mechanism for Yahoo Shopping broken? How could this horrible retailer have a four star rating with 858 ratings? I'm convinced that there is a possibility that many of the 'reviews' for this company could be fake."[52] In response, Price Grabber.com's Business Director of Technology and Entertainment told the press that user reviews posted to PriceGrabber.com are screened by humans and also an automated algorithm to make sure they are not written by one person or organization. "We have very, very low tolerance for game-playing," he said. "If we find out that there's game-playing going on with the reviews, we delete the reviews and suspend the merchant. If it happens again, we terminate the merchant."[53]

Despite such assurances, general periodicals warn readers of the difficulties and dangers of buying from stores that sell on shopping comparison sites but are not known to consumers. Forbes.com noted in discussing the PriceRitePhoto affair that "figuring out when you can trust a seller will soon become more important because comparison sites have begun mixing in auction results, from sites like EBay, and classifieds with new merchandise from established retailers. When you buy via auctions and classified ads, you are generally buying from an individual, as opposed to a business. This makes doing your homework even harder and collecting a refund almost impossible."[54] More generally, a *Buffalo News* article was headlined "Web Sites for Comparison Shopping Don't All Give the Same Prices." It elaborated that "the range of results can vary widely from site to site and the number of merchants that each site search can be vastly different, as well." It added that "trying to save a few bucks by going with the lowest price from an unfamiliar merchant with shaky customer service will only end up costing you in extra aggravation."[55]

For name brands, this is the industrial benefit of fear. The idea that customers fear being disadvantaged and so see the brand as paramount in the new electronic shopping world is great comfort to established retailers and manufacturers, who fear deep discounters and unknown labels. They hope that encouraging desirable customers to associate a brand with trust will encourage loyalty and tolerance for prices that aren't the lowest.

::

Increasingly, merchants realize that they have to reach beyond the web to cultivate trust and loyalty in the customers they want. One reason is that while online purchases are growing strongly, eMarketer's estimate of $70 billion[56] in 2004 amounts to only 2 percent of the Department of Commerce's estimate of a $366 trillion American retail environment.[57] Moreover, marketers note that consumers often use the web, the telephone, catalogs, and physical stores—what retailers call "channels"—to consider their purchases and then make them. A 2004 Forrester Research survey announced that 65 percent of all online consumers had studied an item online and bought that same item offline (from a catalog or a store). The report concluded that "the era of multichannel shopping has arrived" and that "the online consumer died long ago: Shoppers never replaced store buying with the Web; they embraced it as an additional research and buying channel."[58] Forrester consequently urges retailers to "interact with consumers in a coordinated way across every channel—web, phone or face to face."[59] The strategy is to use all the knowledge about customers to enhance cross-channel selling. It means having the ability to present a product to a consumer, charge for it, and arrange its delivery and return via a number of routes—for example, the traditional "bricks-and-mortar" store, a mailed catalog, a mall kiosk, the phone, a website, or an interactive television channel. The impulse, too, is to merge advertising and sales activities across these channels. That isn't surprising; both types of activities have come to mean reaching out to desired consumers through interactivity, targeted tracking, data mining, segment-making, mass customization, and the cultivation of relationships based on those activities. In fact, it is easy to see how the niche-making media activities described in previous chapters about the internet and TV could be meshed with the niche-making retail developments sketched in this one.

The overall industrial logic of marketing communication is to track people who belong to particular niches and to direct customized messages and even media environments to them in as many places as is possible. Executives increasingly believe that careful collection and analysis of individual shoppers' movements across multi-channel "touch points" will gain them competitive advantage over Wal-Mart and its ilk. The desire to implement that strategy has led retailers to build their capacity in cross-channel selling and data integration despite the substantial costs.

Knowing more about individual consumers can also serve as a defense against resistance in which technology encourages price comparison within physical stores. In 2005 both Google and Amazon introduced cell-phone services that allowed customers to compare prices while walking down the aisles of a Best Buy store or a supermarket. Around the same time, an MIT researcher showed the press a "wand" he had developed to aid shoppers concerned to learn about products before they buy them. The wand is a scanner that holds in its memory information on thousands of products. Go down the aisle in a store and brush the universal product code against the wand, and into the wand's window comes information about the item from a variety of standpoints—its price, its durability, even the company's environmental friendliness.

It is hard to see how retailers can stop consumers from bringing such devices into stores, and from the retailer's standpoint they may represent a challenge. But one way to discourage folks from seeing advantages to such devices is to customize the physical shopping situation so that the consumer feels uncomfortable going to another place. That is where cross-platform data come in. What banks do at the branch and on the phone, and what Bloomingdale's does with Klondike on the floor, retailers will be able to do they meet a desired customer.

In fact, with the news media pointing to a frightening commercial environment filled with identity thieves, scams, and hidden costs, retailers are betting that consumers who want stability and trust will step up to be counted. Learning from friends, ads, or the press that media firms and retail stores treat some customers better than others, they will envy the best-treated niches and hope to be catalogued that way by their favorite stores, television networks, websites, and magazines. Assured by those firms of their data's security (whether or not they really believe that), to be treated as special consumers they will give up information about themselves, semi-knowingly consent to being tracked, and often disregard the advice of comparative shopping sites or wands. In the process, they will identify themselves, often biometrically, when they turn on their televisions, go through the doors of their favorite stores, or click into the online versions of those stores.

Marketers argue that this eventuality is the Holy Grail for businesses and consumers. Media and marketing executives get to bring order to a woefully unpredictable environment. Consumers get to be served by marketers who know them intimately. They get to be treated well on the phone by credit-card-firm representatives whose screens tell them how to speak to those consumers as members of a specific customer segment that the firm values. They will receive coupons for products that their supermarket's research of their segment concludes will be among their favorites. Online movie sites will send them rental suggestions based on their previous rentals and on the statistical probability that they will like certain films that rent well in their niche. Database-driven customization may sound wonderfully helpful, but there is a dark side to it. As I will argue in the next chapter, the concerns are individual and societal. And they go to the heart of what it means to be surrounded by ads, articles, programming, product offerings, and retail deals that are offered in the name of trust and good relationships when just below the surface lie distrust, envy, and suspicion on both sides.

8 :: Envy, Suspicion, and the Public Sphere

This book depicts an emerging world in which marketers use sophisticated databases about consumers to create customized appeals, offers, and programs that appear specifically to them across a wide range of media. At present it is not easy to point to clear-cut programs in which media, manufacturing, and retail forces continually use consumer data to merge retail sales and media advertising spaces in ways that customize for various data-driven niches. Cost may be the most important consideration here. A Visible World executive claimed that it would cost $22 million to upgrade Comcast's set-top boxes to accept customization at the household level. Add to that even higher bills for the server speed and memory required to create sites that adjust products, service levels, and discounts to reflect what the companies knows about each consumer and his or her value. Then imagine the added complexity of having these domains interact with one another so that the websites' data are shared with the television provider's data in ways that influence television and web outputs and encourage interactivity with various sales platforms.

But although practical considerations make widespread niche marketing of this sort impossible today, signs are strong that they will eventually come to pass. For one thing, the cost of the technology will continue to plummet. In addition, competition for the right audiences and customers will only get fiercer. In a society roiled by socioeconomic divisions exacerbated by the availability of cheap labor in other countries, manufacturers, retailers, and media firms envious that competitors have tapped valuable niches will want to accelerate their use of technology to do the same. Merchants will try to find, court, and keep those consumers who are worth the effort. Marketers will use advanced electronics to zero in on them, place them into relevant niches, infer the most likely selling points, and then continually promise them those selling points plus the good life if they stay loyal.

This may well result in a marketing-and-media world that changes for every person depending on what niche the marketers put that person in. Consider television viewing. More than a few prognosticators have suggested that in the next decade people will view "television" at home by first turning to intelligent navigators that suggests program lineups for them. It isn't hard to imagine that what shows will be at the top of the list will be determined by analyzing data collected by the television provider (perhaps a cable firm) and data held by marketers. In order to maintain the privacy of individual information, the pooled data will be analyzed only by the computer that serves the programming differentially to different households; the collective information will not be examined or stored by the firms involved.

Different advertisers may place an individual into different niches, and the television provider may have to find ways to allocate priorities to these niches—or some way to reconcile them—when it comes to choosing programming. Based on those categories, the suggested lineup of programs and commercials one household receives may differ from that received in the house next door. In fact, different members of the same household may be treated differently if they are watching different televisions and if they have logged in with individual codes or biometric indicators.

Even choosing a program with the same title may yield different experiences for different people if different advertisers want to reach them. If Sally's television provider has pegged her as a "young unmarried upscale female professional," she may receive a different version of a television movie than would Susan, a "divorced middle-aged middle-class mom with kids." A bank may have bought time to show one kind of commercial to Sally and another to Susan. Product placements may also differ. Ford may have contracted to have the heroine drive a Mustang in the version of a show that Sally's group receives and a Windstar van in the version Susan's group gets. Tiffany may have arranged for an attractive character to mention it by name in Sally's version of the show, while in Susan's version there may be a paid-for mention of another jeweler. If Sally is a member of a Tiffany loyalty club, she may also have the opportunity to click on the expensive brooch that the character in her version wears. Perhaps she can learn more about that brooch online, and perhaps she can receive a gift for going to a Tiffany store in a nearby mall. This may make Sally feel good, but only for a day. By chance, she may learn that a wealthier woman living nearby has received a more expensive gift for visiting the store and has been invited to a special fashion show. Sally may then begin to wonder what Tiffany knows about her.

Because no two households or individuals are likely to have the same companies trying to reach them, the programming that two households or individuals are likely to see will be different. News networks may suggest—and even stream—different stories to them on the basis of what the networks and the advertisers know about them and think they will like. What coupons your set-top printer will run off may depend on what niche the manufacturers or retailers who want to reach you have put you in. Perhaps those in the best niches will receive the best coupons, and those coupons will be nontransferable. What offers you receive may depend on a number linked to a biometric measure (or, for a household, a set of measures) that you gave freely when you signed up for that company's "best customer" card. In these still-imaginary scenarios, niche-customized activities will take place, and will interrelate, in media and retail outlets that you use when you use them. The industrial logic of the marketing-and-media system points in exactly this direction. The message of this emerging system to consumers is muted today, but it is growing louder. That message is: "If you want the best deals you can get, sign up to be a 'best customer.' You will be required to contribute information about yourself and your household that will help us understand you better. You don't have to do it, but if you don't do it you may miss out on many opportunities. In fact, you may, literally and figuratively, not count."

::

As I noted in chapter 2, the new approach to consumers marks a profound change in American culture. Broadly speaking, the past 150 years saw what might be called the democratization of shopping. Beginning in the mid 1800s, department stores such as Stewart's in New York and Wanamaker's in Philadelphia moved away from haggling and began to display goods and uniform prices for all to see. One part of the motive was self-interest: With a wide variety of merchandise and a large number of employees, the store owners didn't trust their clerks to bargain well with customers. But the result was a fairly egalitarian and transparent marketplace, with products and prices that all could see.

Reliance on open and even-handed dealing is central to American capitalism's public image. It is not always practiced, as antitrust suits and many consumers' complaints attest. Nevertheless, the Securities and Exchange Commission and the Federal Trade Commission were established, in part, to aim for it.

In chapters 3–5, I described the shift away from the old verities of business and the emergence of a new industrial logic. The activities that flow from this logic don't result from any single-minded conspiracy against the public. They are emerging among retailing, marketing, and media firms as they work, separately and together, to meet serious challenges. Though the challenges differ, they overlap and influence one another. Media executives worry about fragmentation of audiences due to splintering of media channels. Practitioners of marketing communication worry about the best ways to reach people through those splintered channels. Media personnel and marketing-communication workers agonize about the increasing ability of consumers to avoid advertising messages altogether and about whether product-integration activities are persuasive alternatives. Retailers who advertise worry about these things too, and also about the increasing pressure they feel from Wal-Mart and other low-price retailers and from the growing number of internet competitors.

Executives in the marketing-and-media system search for solutions to their problems by talking to one another, by reading the trade press to learn about innovations, by going to conferences, and by trying new operations. Consultancy firms play a particularly important role in disclosing what other firms are doing and by proselytizing for alleged best practices. The solution they see is a marketing-and-media system whose very survival is tied to the gathering of information about consumers. The database marketing in which that proposition is being implemented typically includes six activities:

Screening for appropriateness Using information they have collected or bought, marketing or media firms make judgments as to whether they want particular individuals as customers.

Targeted tracking Marketing or media firms follow actual or potential customers' marketing and/or media activities to learn the consumers' interests and to decide what materials to offer them.

Data mining To learn about the characteristics that will draw and keep the interest of actual or potential customers, marketing and media firms explore the data they have collected by tracking them, registering them, or purchasing information about them.

Interactivity To draw individuals toward their products, marketing and media firms encourage actual or potential customers to interrogate the firm's virtual or actual representatives in the process of evaluating and choosing the products they want.

Mass customization To draw individuals toward their products, marketing and media firms use what they have learned from targeted tracking and interactivity to offer tailored choices to customers based on specific niches in which they have placed those customers.

Cultivation of relationships When the information gathered shows that customers fit into niches that marketing and media firms desire, the firms initiate actions (including mass customization and interactivity) aimed at establishing bonds with those customers so that they will keep coming back.

The sequence of these activities may differ from the order in which I have just presented them. The activities inform one another, continually building on the data presented. For example, companies may be continually adjusting the niches in which they place customers on the basis of what they learn from regularly tracking them, mining their data, and evaluating their responses to particular mass customizations. In addition, the specific nature of these activities may vary among retailers, among media firms, and among media forms. The internet has become a test bed for these activities. Television, physical stores, and other marketing venues such as video games don't yet have the technological capabilities for the kinds of targeted tracking, interactivity, and mass customization that are available via broadband.

Yet even the web is immature, as marketing and media executives see it. Their new industrial logic leads them to work toward a world in which databases rule. It is a world in which biometric data recognition provides executives with a secure sense of who the entering consumer is, in which programming, product offerings, and price discounts are customized instantly on the basis of a customer's history and niche identification, and in which the entire process reinforces the consumer's connection to the relationship while adding information about the encounter to the data set so that the next encounter will be more profitable.

Note that this is subtly different from the twentieth-century model, in which storekeepers stocked certain brands because they sold well, thus pleasing customers and making money at the same time. Though that certainly continues, the new goal is to make money by identifying individuals who fit "best customer" profiles and then reinforcing their purchases for reasons and in ways that are hidden from them.

::

An emerging awareness that surreptitious activities are taking place is beginning to drive what might be called a new culture of suspicion and envy. Americans have learned to take collectively displayed products with posted prices for granted. As automobile marketers know well, when consumers have to negotiate over price—most notably at dealerships—they tend to see that as an unusual, nerve-wracking experience. A belief in marketplace openness applies to the web, as well. The 2005 Annenberg Center survey discussed in chapter 7 found that most people believe that it is unfair for a retail website not to show them the same products that other visitors to the sites see.

It should not be surprising, then, that the scaffolding of this system is shaken if a retailer changes its offerings to individual consumers (or communicates with them differently) because of niche-based information about the consumers that the consumers don't know of, or that they suspect but can't verify. The 2005 Annenberg Center shopping survey suggests that, at this point in the development of the system, American consumers seem to disagree most about database marketing when they hear or believe that discrimination in pricing is taking place. In chapter 6, I mentioned a number of clear cases of price discrimination based on individual tracking that commonly take place at banks, at supermarkets, and at other offline retailers. When online price discrimination is guided by the tracking of people's behaviors or backgrounds, it is extraordinarily hard to verify that such discrimination occurs, why any particular offer is made, or how a vendor is evaluating any given customer. Web merchants don't have to tell anyone how they operate, so generally they don't.

As I noted in chapter 7, airlines' pricing structures and Amazon's variable pricing of CDs are among the developments that are leading internet-savvy consumers to suspect that price customization may be taking place. Amazon apologized for what it said was merely a "price test," and the airlines say the changing rates are merely the result of a necessarily complex pricing structure. Both explanations may be accurate. Yet it is hard for any dispassionate observer to believe that price customization is not going on when associates from the influential consulting firm McKinsey trumpet its importance by asserting in a 2004 *Harvard Business Review* article that online companies are missing out on a "big opportunity" if they are not tracking customers and adjusting prices accordingly.

Anecdotal evidence suggests that many savvy internet users don't believe Amazon's and the airlines' assurances:

• In my undergraduate seminar on "Spam and Society," the discussion veered a bit off the topic. One student asserted confidently that airlines' web sites offer first-time users prices lower than those offered to returning customers. Most of the others immediately agreed. The motive, they said, was to suck in potential buyers so that when they returned the airline could quietly raise prices.

• An article I wrote on database marketing for the *Washington Post* led several people to email me reports of what they were sure were instances of price discrimination. One such report went as follows: "I was buying a web camera and did some research on the internet. Among the web sites I visited were those of the manufacturers, Creative and Logicon, and two retailers that have both an internet store and bricks-and-mortar stores, Circuit City and Best Buy. I assume all of these web sites left their cookies behind. I selected two models I was interested in, one by Creative and one by Logicon. Both models were at the same price, $49.99, at both Circuit City and Best Buy. I checked the store stock online at the closest store to my house, a Best Buy in Pentagon City, VA. When I went into the Best Buy store, the shelf price of the Logicon web cam was $49.99, but that of the Creative web cam was $51.99. I asked a store clerk to price scan the Creative web cam, and it came back at $51.99 also. I told him I had just looked at their web site within the hour, and the price there was $49.99. I asked if we could visit the web site from within the store. He took me to a computer and we visited the web site from there. The screens looked identical to what I had seen from home, except the price accessing the web site from the store was $51.99. I mentioned this price manipulation topic to the store clerk, and he seemed totally ignorant of the concept. Who knows if he was telling the truth, or if he was just 'playing dumb' by company policy? I asked to speak to a manager about the issue. Luckily one was close by, and she was wearing a name tag that said 'General Manager,' for what it is worth. I got into what was a very short discussion of the price manipulation with her, and almost immediately she told the sales clerk to give me a price match to BestBuy.Com. I don't know if she was just extremely busy, or if she just wanted to silence me as quickly as possible so no one standing around would overhear the discussion. I'm sure their average customer would be very troubled hearing about such things. After I went home, I again checked on the price of this Creative web cam on the Best Buy web site. It was still at $49.99 (after clearing the cache to make sure I got a reload of the web pages). This was within two hours of my initial price check on their web site. For what it's worth, the web cam is still listed at $49.99 on their web site today, though I have no idea what the store price is now. I realize a lot of stores charge a different amount for an item at their online stores versus their bricks-and-mortar stores, and that's their prerogative. But the fact that the access to the online store from a computer within one of their bricks-and-mortar stores gives a different price than from within a customer's home is very disturbing."[1]

• A few of the aforementioned emailers were less indignant than envious that others were getting better deals. In a Washington Post chat room dedicated to database marketing, a woman wrote that "the heart of the matter" is "finding strategies one might use to game the system in their favor." "How," she asked, "can a user tailor [her] behaviors to maximize the likelihood of always getting the best price?"

Such comments point to new kinds of distrust, suspicion, envy, and accommodation that are emerging around retailing and the advertising and media connected to it. But concern about price discrimination is only the leading edge of a so-far-small percentage of the population that knows what is taking place. At this point, the industrial logic of the new database age is not centered on, or even primarily concerned with, the customizing of prices to lure desirable customers. As I noted in chapter 6, in a hypercompetitive environment in which trying to beat Wal-Mart and Costco on price is all but impossible, department stores and supermarkets compete by trying to hook the right customers. Operating on the financial industry's premise that about 20 percent of the customers bring in 80 percent of the profits, they try to identify who belongs in that 20 percent—who will spend money and come back to spend more. And they often try to get rid of those who hold out for bargains or return too many purchases.

As the activities described in this book take hold of the retailing and media establishment in the coming decades, the population at large will learn that consumers daily confront not only customized prices but also customized advertisements and media materials. Sometimes firms will let consumers know about characteristics that make them count. Microsoft, for example, announced in 2005 that it was instituting a new online advertising program in which targeted individuals would sometimes "be able to mouse over the ads that are displayed, learn why the ads were targeted to them, have some input in what ads they will see, and offer feedback about the ad, that will then be made available to the advertiser."[2] When firms are not disposed toward this type of transparency, news reports and gossip will suggest to consumers that they are involved in customized relationships with merchants and media firms based on what those companies know about them. The consumers will be not always be sure when discrimination is taking place, but they will suspect it. They may be angry about it, but often, much like the writer in the *Washington Post* chat room, they mostly will envy others who they think may be getting more interesting announcements or better deals. And they may well try to tailor their behaviors so as to provide Microsoft or their favorite store with the kind of information that will maximize their likelihood of getting the best deals.

::

Privacy advocates justifiably worry that personal information taken secretly from individuals may be used in ways that the individuals would not want. Marketers are learning, however, that they can get around the privacy bugaboo. They simply ask desired customers for personal information in return for promising to engage in beneficial, trust-building relationships with them. The companies follow the Federal Trade Commission's guidelines in regard to giving their customers data security, choice about whether or not to give the information, notice of basic ways the information will be used, and access to the information the customers have provided.

One problem with this alleged openness regarding private information, as I noted in chapters 4 and 7 , is that it isn't really open. Customers don't fully understand the implications of giving up data, and they can't really gain access to all the data the companies have gathered about them. (They just know what data they supplied.) Moreover, even when customers feel uneasy giving out personal information, they often believe that they *must* enter into database relationships to get good deals.

But there is more concern about database marketing than there is about the secret use of data for harmful ends. Oscar Gandy and other writers have pointed to the unfairness of using databases to put people into categories with which they might not agree for purpose which they might not agree.[3] For decades, marketing and media firms learned as much as they could about social groups (women, baby boomers, rich people, African-Americans, and so on) and then tried to target people they thought were members of those groups. The emerging process is almost the opposite: They learn enormous amounts about individuals, consign them to various niches, and then determine whether and how they want to deal with them.

Only a few marketing practitioners seem to have worried about the ethical problems associated with such activities. One example is a 1999 guide to customer-relationship marketing by Paul Gamble, Merlin Stone, and Neil Woodcock. In this book (titled *Up Close and Personal?*), Gamble et al. use the term "moral maze" to describe customer-relationship marketing's need to "make regular decisions about the worth of other people." Nevertheless, they conclude that this differentiating must be done. "Organizations," they write, are "increasingly keen to use their customer databases to develop profiles of good and bad customers so that they can categorize new or potential customers before entering into a relationship."[4]

When customers know that media firms and retailers they frequent are using their data but aren't sure how they are using it, it is understandable and reasonable that the customers may be wary of their status in those relationships. Yet the development of a database-driven culture of suspicion toward marketplace and media activities should also ring alarm bells among those who care about a healthy public sphere. The marketing-and-media system is a crucial contributor to publicly shared stories and discussions. People rely on both the market and the media to learn what the trends are and where they stand when it comes to fashion, politics, media preferences, and other aspects of social position.

Having the option to share the same marketplace of goods and ideas has become a central proposition of equality in the United States. Yet in the new world of database-driven marketing individuals will consistently worry that the items and stories they receive are different from those others are seeing, for reasons they don't understand. They will find it difficult to decipher what marketers and media think of their social status. The envy and suspicion that result from that insecurity will generate new worries about how to reveal oneself in public when doing so may reveal information that may be inserted into databases and may then have unknown consequences for one's social choices. People might, for example, choose to say or do certain things because doing those things seems likely to get them placed on lists that will get them better coupons for more upscale products, get advertisers to support their pay-per-view cable movies, or get the magazines they receive to include ad-supported inserts on certain topics.

This type of concern already exists among those who understand the new flow of information. Employers often "Google" people before hiring them, and individuals often "Google" others before going on blind dates with them. Nasty claims linked to a searched name or unfortunate comments made on the web years ago may affect one's chances of employment or one's personal life. Multiply such worries to include data in the entire retail, marketing, and media system. Include the information about individuals, their interests, and their social connections that companies might gather from search engines and from "consumer-generated media" such as blogs, photo-sharing locations, and "tag" sites. The resulting data trail will make it impossible for many people to feel a sense of basic trust in the openness that is important to the vigorous sharing of ideas, opinions, and arguments within and across segments of society—that is, to a healthy public sphere.

Taken to their logical ends, recent developments in marketing and media raise questions about the nature of media and sharing that did not come up in the past. What happens to social conversation when people who watch the nightly news or a particular evening drama can no longer assume that they receive the same material that others receive? What happens to a person's sense of predictability when she goes online to buy something on a website at the suggestion of a friend and finds that to get to the items her friend saw on the home page she must navigate through many screens? Who will create opportunities for various social groups to talk across divisions and share experiences when major marketing and media firms solidify social division by separating people into data-driven niches with news and entertainment aimed primarily at reinforcing their sense of selves so they will buy?

It is particularly difficult to answer these questions because at any point in time consumers will not know that unwanted customization is happening, or how or why it is being done. Much of the time, people may even feel good about the attention they are getting in their favorite stores, from their cable system, and from their internet provider. On the one hand, marketing and media practitioners need to get consumers to trust them and their brands in order attract their business and to persuade them to provide information. On the other hand, many of the same practitioners are deeply fearful that consumers will use new technologies to refuse their ads, reject their products, or find lower prices.

The fear leads media and marketing firms to collect and exploit data on customers in more ways than they want the customers to know about, so they can counteract the customers' power by discovering how to keep the their attention (and, of course, so they can optimize their profits). From the marketing and media practitioners' standpoint, the goal is a worthy one in a brutally competitive environment. But with respect to consumers' hope that trusting marketers with their data would be accompanied by marketers' openness about the ways they use consumer information, what has been happening instead is that marketers have been rewarding consumers' trust in them with an undermining of that trust through the surreptitious ways they use information.

American consumers today have only a glimmer that "niche envy" by retailers and media firms is leading them to furiously collect, analyze, and try to use data about just about anyone who comes through their doors. As I noted in chapter 7, The Annenberg Public Policy Center's 2003 and 2005 national telephone surveys of American adult internet users suggest that, although the great majority gave up personal information to get frequent-shopper cards and believe that companies have the ability to track their web activities, they have little clue about data mining or how companies use the data they gather. Moreover, not only do they object to most forms of behavioral targeting, customized product displays, and price discrimination (on the web and off); they think behavioral targeting and price customization are illegal. Nevertheless, news reports of identity theft and information abuse may contribute to a nagging suspicion that companies may be using their data without their permission in ways they don't want. Fully 79 percent agreed with the statement "I am nervous about websites having information about me," and only 17 percent agreed with the statement "what companies know about me won't hurt me." Moreover, only 35 percent agreed with the statement "I trust the U.S. government to protect consumers from marketers who misuse their information."[5]

Already, then, we are seeing the evolution of a culture of suspicion mixed with envy. Small incidents reflect the trend. A friend of mine phoned Citibank about his credit card and got a representative from India who wasn't able to understand his problem. The customer had never before gotten an "outsourced" operator. Knowing something about databases, he concluded that Citibank had pushed him down a notch in status, reserving the U.S.-based representatives for better customers. In reality, he knows nothing about how Citibank operates, and he has no idea what it thinks about his status. It is because he is internet-savvy that he is suspicious that information may be used against him without his knowing it.

From airlines to supermarkets, from banks to web sites, American consumers increasingly believe they are being spied on and manipulated. The idea used to be that a consumer could shop around, compare goods and prices, and make a smart choice. But now the reverse is also true: The vendor gathers information about a person and decides whether it can profit from his or her loyalty and habits. A customer may not feel comfortable giving up personal information when applying for a frequent-shopper card, but many people do it because they feel powerless to resist and they want to make sure they get deals at least as good as those that others get.

All this may make sense for retailers, but for us customers it can feel as if our simple corner store is turning into a Marrakech bazaar—except that the merchant has been analyzing our diaries while we negotiate blindfolded, behind a curtain, through a translator. "My money is as good as anyone else's" has been a common American expression, but that may no longer be true. Children born in the coming decade may know only a society in which much of the advertising, product offers, news, and entertainment that they receive is customized on the basis of information that companies collect and store about them. They and their offspring will try to beat the system so as to get the best deals on products, news, and entertainment, all the while assuming that they don't know all the rules of the marketing-and-media system or what the system knows about them. They will be mistrustful and cynical. It is a disconcerting scenario, driven by basic changes and challenges in the American economy.

The phenomenon probably can't be stopped, but there may be ways to slow it down and weaken it. The overall goal should be to insist on openness in customer-company relationships when it comes to the use of people's information. Jeff Woods, an analyst with Gartner Research, has written about the social discussion that he feels should take place about the use of radio-frequency identification in retailing. His comments are applicable to the entire spectrum of activities associated with consumer databases. "We are setting up a framework and architecture for the next twenty or thirty years of commercial activity," Woods told *Risk Management*. "A legitimate question is whether we are setting up a framework and architecture in which we want to live."[6]

The social discussion should be accompanied by small and large social policy initiatives that encourage openness and discussion about database trends in media and marketing. Here are three:

• The Federal Trade Commission should require websites to replace the label "Privacy Policy" with "Using Your Information."

We found in 2005 that 75 percent of internet-using adults did not know the correct response ("False") to the statement "When a website has a privacy policy, it means the site will not share my information with other websites and companies." For many people, then, "Privacy Policy" is deceptive; they assume that it indicates protection for them. Calling the same item "Using Your Information" will likely go far toward reversing the broad public misconception that the mere presence of a privacy policy automatically means that the firm will not share the person's information with other websites and companies.

• School systems should develop curricula that tightly integrate consumer education and media literacy for students in all grades.

We found that a high level of general schooling doesn't necessarily mean that one is well informed about the laws and practices regarding behavioral targeting and price discrimination or about where people can turn for help if marketplace information is used illegally. We conclude that specific consumer education linked to media literacy is needed, in addition to general schooling, to improve the public's understanding of market practices.

Consumer education (often considered a part of economic or financial education) varies dramatically from state to state. Several non-profit organizations, including the Jump$tart Coalition for Personal Financial Literacy and the National Council on Economic Education, have as their goal the financial competency of America's young people. According to Jump$tart, in early 2004 only 15 percent of American high school graduates had taken a course covering the basics of personal finance.[7]

There is a growing awareness of the need to make financial education a priority both at the federal level and at the state level. The 2002 education bill commonly called the No Child Left Behind Act includes an Excellence in Economic Education (EEE) program to promote economic, financial, and consumer education from kindergarten through the twelfth grade. In July 2004, the U.S. Department of Education granted its first EEE award ($1.48 million) to the National Council on Economic Education.[8] Though advocates of financial education for youngsters applaud the grant, they also point out that the amount awarded is small relative to the work that is to be carried out.

If consumer education has little visibility in the schools, media education is virtually nonexistent. Educators typically justify the lack of attention by saying that they have enough difficulty covering the standard curriculum; they consider media education a luxury. But the developments that motivated our survey should underscore one reason why media literacy is a necessity rather than a luxury. More and more, media vehicles are becoming integral to the selling environment. Computers that display advertisements (some of them interactive) are showing up on supermarket shopping carts. In the checkout areas of all sorts of retailers, discount coupons are printed selectively on the basis of data accumulated during previous visits or bought from brokers. Websites use a myriad of data-collection approaches that have consequences for the advertisements people see, the products they encounter, and the prices they pay. These techniques and others are changing the shopping and media landscapes. Educators must integrate an understanding of media and of marketing into the curriculum so that future students will not be as ignorant, fearful, and distrustful with regard to trends in the marketplace as today's adults are.

• The federal government should require retailers to disclose specifically what data they have collected about individual customers, and to disclose when and how they use the data to influence interactions with them.

In one of the saddest findings of our survey, 81 percent of respondents disagreed with the statement "What companies know about me won't hurt me." This basic, widespread concern that businesses' collection of information about individuals can cause them harm ramified through the interviews. It showed up most prominently in our several attempts to tap into people's attitudes toward different forms of price discrimination. Perhaps sometimes to the point of naiveté, the nationally representative sample of internet-using adults insisted on fairness in pricing. Fully 91 percent thought it wrong for a supermarket to charge people differently for the same products within an hour, 87 percent said the same about online stores, and 84 percent said that websites should be required to let customers know if they vary the prices of items within an hour.

::

Clearly, people are begging for transparency in their relationships with marketers. It may well be that internet-using adult Americans, if informed about now-surreptitious price discrimination activities that affect them, would still view the practices as unfair. But they believe it is their right to know. Perhaps in an environment of greater trust and openness certain kinds of preferential dealings would be acceptable, just as publicly announced price preferences for senior citizens are accepted today.

Actions by the federal government are critical to establishing an atmosphere of marketplace transparency and trust. The broad disagreement we found with the statement that the U.S. government will protect consumers from marketers who misuse their information indicates there is much that public officials must do to regain the public's trust. It also suggests the connection between people's attitudes as consumers and their roles as citizens. A well-developed, critically informed understanding of how the media and commerce now work together can have favorable consequences for the ways people view important social institutions, as well as for what they know about themselves, about their neighbors, and about the goods they buy.

Notes

Chapter 1

1. Richard Barlow, "How to Court Various Target Markets," *Marketing News*, October 9, 2000, p. 22.

2. Christine O'Dwyer, review of Harvey Thompson's *Who Stole My Customer?* O'Dwyer's PR Services Report, May 2004.

3. J. Walker Smith and Craig Wood, "Difference Matters," *Direct*, August 1, 2003.

4. Jon Swartz and Byron Achoido, "AOL Breach Gives Spam Fight a Twist," *USA Today*, June 25, 2004.

5. In "phishing," thieves dupe consumers into entering personal data on counterfeit banking and e-commerce websites. See David McGuire, "Bush Signs Identity Theft Bill," Washingtonpost.com, July 15, 2004.

6. Direct marketing to the very wealthy sometimes is done by means of letters and even personal visits.

7. See Daniel Boorstin, *The Americans* (Random House, 1958).

8. Susan Matt discusses the early history of this perspective in *Keeping Up with the Joneses* (University of Pennsylvania Press, 2003).

9. I do not capitalize 'internet'.

10. Tony Cram, "Marketing Society—How to Care for Customers Who Count the Most," *Marketing*, July 12, 2001, p. 20.

11. Richard H. Levey, "Bloomingdale's Goes for the Best," *Direct*, January 1, 2004, p. 1.

12. See, for example, Robert Harrow, *No Place to Hide* (Free Press, 2005); Simson Garfinkel, *Database Nation* (O'Reilly, 2000); Oscar Gandy, *The Panoptic Sort* (Westview, 1993); Daniel Solove, *Digital Person* (New York University Press, 2004);

Daniel Solove and Marc Rotenberg, *Information Privacy Law* (Aspen, 2003); Peter Swire, *None of Your Business* (Brookings Institution Press, 1998).

13. See, for example, Paul Starr, *The Social Transformation of American Medicine* (Basic Books, 1982).

14. Martin Mayer, *Madison Avenue, USA* (Harper, 1958),

15. James Playsted Wood, *The Story of Advertising* (Ronald, 1958).

16. See Stuart Ewen, *PR! A Social History of Spin* (Basic Books, 1996).

17. Matthew McAllister, *The Commercialization of American Culture* (Sage, 1996).

18. Thomas Frank, *One Market under God* (Doubleday, 2000).

19. See, for example, Andrew Johnson, "Radio Stations Tune In to Listeners," *Washington Times*, February 28, 2005.

20. See, for example, Molnar's comments in Bill Pennington, "Reading, Writing and Corporate Sponsorship," *New York Times*, May 24, 2005, and in an editorial titled "We're for Sale," *Pittsburgh Post-Gazette*, May 24, 2005.

21. Jean Halliday, "The Gospel According to Martin and John," *Advertising Age*, May 16, 2005, p. 1.

22. Scott Donaton, "It's Time to Take a Fresh Look at the Definition of Marketing," *Advertising Age*, February 23, 2004, p. 16.

23. Bradley Johnson, "Cracks in the Foundation: Why the Very Currencies the Industry Depends On Are Dated and Inadequate," *Advertising Age*, December 8, 2003, p. 1.

24. Devin Leonard, "Nightmare on Madison Avenue," *Fortune*, June 2004, p. 92.

25. David Tice of Knowledge Networks/SRI, as quoted in Joe Mandese, "'Big Bang' Rattles TV Business, Pace of Technology Accelerates," *Mediapost Daily News*, June 30, 2004.

26. Laura Mazur, "TV Is Being Left Behind by Hunt to Improve CRM," *Marketing*, February 26, 2004, p. 18.

27. Francis Fukuyama, *Trust* (Penguin, 1996), p. 26.

28. Anthony Giddens, *The Consequences of Modernity* (Stanford University Press, 1990)

29. O. Renn, C. Jaeger, E. Rosa, and T. Webler, "The Rational Actor Paradigm in Risk Theories: Analysis and Critique," in *Risk in the Modern Age*, ed. M. Cohen (Palgrave Macmillan, 2000).

Chapter 2

1. Caroline Marshall, "Backbite," *Campaign*, October 31, 1997.

2. Winston Fletcher, "Soap Box: The Notion of Ad Wastage Is Just Meaningless Pap," *Marketing*, February 29, 1996.

3. Edwin Emery and Michael Emery, *The Press in America*, fifth edition (Prentice-Hall, 1984), pp. 199–200.

4. James Playsted Wood, *The Story of Advertising* (Ronald, 1958), pp. 45–69, 186.

5. Daniel Boorstin, *The Americans* (Random House, 1958).

6. Quoted in Wood, *Story of Advertising*, p. 234.

7. Wood, *Story of Advertising*, p. 232.

8. Ibid., p. 207.

9. Frank Rowsome Jr., *They Laughed When I Sat Down* (Bonanza Books, 1959), p. 62.

10. Ibid., pp. 134–135.

11. Susan J. Matt, *Keeping Up with the Joneses* (University of Pennsylvania Press, 2003), p. 13.

12. Ibid., p. 26.

13. Edward Bok, "At Home with the Editor," *Ladies Home Journal* 8 (September 1891), p. 10

14. Matt, *Keeping Up with the Joneses*, p. 38.

15. Quoted in ibid., p. 42.

16. Matt, *Keeping Up with the Joneses*, p. 2. She cites others.

17. Wood, *Story of Advertising*, p. 343.

18. Quoted in ibid., p. 213.

19. Wood, *Story of Advertising*, p. 405.

20. Ibid., p. 412.

21. Stuart Ewen, *Captains of Consciousness* (McGraw-Hill, 1976), p. 35.

22. Ibid., p. 38.

23. Robert S. Lynd, "The Consumer Becomes a Problem," in *The Ultimate Consumer*, ed. J. Brainard (American Academy of Political and Social Science, 1934), p. 6.

24. Cited in Wood, *Story of Advertising*, p. 435.

25. Neil Borden, *The Economic Effects of Advertising* (Prentice-Hall, 1942), p. 881.

26. See Ellen P. Goodman, "Media Policy Out of the Box," *Berkeley Technology Law Journal* 19 (2004), no. 4: 1390–1472; J. Rothenberg, "Consumer Sovereignty and the Economics of TV programming." *Studies in Public Communication* 4 (1962): 45–54; C. Edwin Baker, "Giving the Audience What It Wants," *Ohio State Law Journal* 58 (1997): 316–338.

27. Ibid., p. 8.

28. Lawrence Van Gelder, "Milton Berle, TV's First Star as 'Uncle Miltie,' Dies at 93," *New York Times*, March 28, 2002.

29. See Christopher Sterling and John Kittross, *Stay Tuned*, second edition (Wadsworth, 1978), p. 657.

30. Joe Mandese, "Death Knell for Demo?" *Advertising Age*, July 25, 1994, p. S-2.

31. See Joseph Turow, *Breaking Up America* (University of Chicago Press, 1997), pp. 55–89.

32. See, for example, the entry on DeForest in Wikipedia (http://en.wikipedia.org/wiki).

33. Lee DeForest, "Dr. DeForest Designs the Anti-Ad," *Radio News*, September, 1930, pp. 215, 285.

34. See, for example, Craig Reiss, "Fast-Forward Ads Deliver," *Advertising Age*, October 27, 1986, p. 3; Ernest Schell, "Learning to Live with VCR," *Advertising Age*, April 3, 1989, p. 3.

35. See, for example, Joe Mandese, "ABC Test Squeezes Clutter," *Advertising Age*, August 17, 1992, p. 46; Chuck Ross, "Study: Commercials Battle All-Time High TV Clutter," *Advertising Age*, December 14, 1998.

36. Evan Hansen and Stefanie Olsen, "Spam: It's More Than Bulk E-Mail," CNET News.com, October 8, 2002.

37. Ibid.; Saul Hansell, "AOL Providing Software to Customers to Block Pop-Ups," *New York Times*, March 12, 2003.

38. http://www.aol.com/info/bulkemail.html, accessed March 21, 2003.

39. Todd R. Weiss, "AOL Ramps Up Fight against Spam for Users," Computerworld online, February 20, 2003.

40. Jeff Kagan, quoted in Weiss, "AOL Ramps Up Fight against Spam."

41. Paul Graham, "A Plan for Spam," in Graham, *Hackers and Painters* (O'Reilly, 2004).

42. Bradley Johnson., "TiVo, ReplayTV Vie for Uncertain Prize," *Advertising Age*, November 4, 2002, p.12.

43. Wayne Friedman, "Panel Discussions: Marketing in a TiVo World," *Advertising Age*, February 10, 2003, p. 13.

44. Staci D. Kramer, "Content's King," *Cable World*, April 29, 2004.

45. Staci D. Kramer, "Kamie Kellner Isn't Against PVRs," *Cable World*, February 24, 2003.

46. Ted Johnson, "TiVo-lution," *Variety VLife*, June 7–13, 2004, p. 52.

47. Stuart Elliot, "Marketing Research Back in Style," *New York Times*, April 18, 2005.

48. Friedman, "Panel Discussions: Marketing in a TiVo World," p. 13.

Chapter 3

1. Jean Halliday and Claire Atkinson, "Pontiac Gets Major Mileage Out of $8 Million 'Oprah' Deal," *Advertising Age*, September 20, 2004, p. 12; Laurel Wentz, "At Cannes, the Lions Say 'Grrr,'" *Advertising Age*, June 27, 2005, p. 1.

2. "Find Formula for a Stronger Week" (editorial), *Advertising Age*, September 27, 2004, p. 32.

3. Richard Edelman, quoted in Matthew Cramer, "Madison Ave. Takes NY," *Advertising Age*, September 27, 2004, p. 22.

4. Aaron Walton, quoted in Halliday and Atkinson, "Pontiac Gets Major Mileage Out of $8 Million 'Oprah' Deal."

5. G6 sales did not rise commensurate with these numbers. Some analysts blamed the campaign but most seemed to agree that one PR success could not overcome a tepid brand image. As the car information website Edmunds.com noted: "This was truly a big score for Pontiac's publicity team because everyone in America heard about it. But chances are, upon a test-drive, you weren't all that impressed, based upon the sales figures." ("Oprah's Pontiac G6 Giveaway Wins Big at Cannes," Edmunds.com, posted June 27, 2005)

6. Halliday and Atkinson, "Pontiac Gets Major Mileage Out of $8 Million 'Oprah' Deal."

7. Nat Ross, *A History of Direct Marketing* (Direct Marketing Association, 1990), p. 5.

8. Rowsome, *They Laughed When I Sat Down*, p. 45.

9. Ibid.

10. Wood, *Story of Advertising*, pp. 254–255.

11. T. J. Jackson Lears, *Fables of Abundance* (Basic Books, 1994), p. 24.

12. Ibid., p. 142.

13. Ibid.

14. Randall Rothenberg, "Marketing's 'Borders' Blurred by Product Placement Revival," *Advertising Age*, September 10, 2001, p. 24.

15. Kerry Seagrave, *Product Placement in Hollywood Films* (McFarland, 2004), p. 65.

16. Ibid., pp. 43–45

17. Ibid.

18. Ibid.

19. Ibid.

20. Ibid., p. 47.

21. Samuel Warren and Louis Brandeis, "The Right to Privacy," *Harvard Law Review* 5 (1890): 193–220.

22. Quoted in Wood, *Story of Advertising*, p. 408.

23. Quoted in Wood, *Story of Advertising*, pp. 408–409.

24. "Product Payola Loading Airwarves Down; Pluggers Trip over Each Other," *Variety*, February 7, 1951: 1; "Product Plug Chiseling on Web Airers Now a Big Payola Operation," *Variety*, April 11, 1951: 1.

25. See Erik Barnouw, *Tube of Plenty* (Oxford University Press, 1975), pp. 243–248.

26. Ibid., p. 263.

27. See, for example, "75 Years of Ideas," *Advertising Age*, January 24, 2005, p. 16.

28. Jennifer Boeth, Jennifer Greenstein, Rebecca Hirschfield, and Kate Stohr, "Go Figure," *In Style*, fall 1997; "Celebrities Can Help Make Sunglasses a Hot Item," *Toronto Star*, July 30, 1988.

29. Seagrave, *Product Placement*, p. 142.

30. Ibid., p. 177.

31. Ibid.

32. David Kalish, "Now Showing: Products!" *Marketing & Media Decisions* 23 (1988), August: 28–29.

33. Marcy Magiera and Alison Fahey (with Jon Lafayette), "Coke Nabs Movies; 2 Big Deals Leave Entertainment King Pepsi Flat," *Advertising Age*, March 16, 1992.

34. Wayne Friedman, "Eagle-Eye Marketers Find Right Spot, Right Time," *Advertising Age*, January 22, 2001, p. S2.

35. Jack Meyers, quoted in David Verklin, "Merely More Channels on the Tube," *Advertising Age*, November 28, 1988, p. S-7.

36. Jim Benson, "Trouble Corralling the Grazers," *Advertising Age*, November 28, 1988, p. S-4.

37. "Tartikoff Sees Long-Term Movie Tie-Ins," *Advertising Age*, November 11, 1991, p. 8.

38. "CSPI Calls for Movie Subtitles Identifying Products," *Broadcasting* 116 (1989), April 3: 57–58.

39. Seagrave, *Product Placement*, p. 179.

40. Marcy Magiera, "Product Placement Group Is Formed," *Advertising Age*, September 16, 1991, p. 46.

41. Marcy Magiera, "As If Having a Hit TV Show Isn't Enough," *Advertising Age*, August 10, 1992, p. 3.

42. "Get Some Action," *Ad Day*, January 21, 1982, p. 3.

43. Ibid.

44. Ibid.

45. Stan Rapp and Tom Collins, *The Great Marketing Turnaround* (Prentice-Hall, 1990), p. 110.

46. Michael J. Weiss, *The Clustering of America* (Harper and Row, 1988), p. 2.

47. J. Fred MacDonald, "The Clutter in the Mailbox Ain't No Junk,'" *Advertising Age*, January 17, 1983, p. M-32.

48. Jo Anne Pagnetti, "Sales Sprout from the Seeds of Segmentation," *Advertising Age*, January 17, 1983, p. M-9.

49. Joseph Turow, *Breaking Up America* (University of Chicago Press, 1997), p. 129.

50. See MacDonald, "The Clutter in the Mailbox Ain't No Junk."

51. "Direct Marketing Flow Chart," *Direct Marketing*, May 1985, p. 21; December 1987, p. 25; December 1990, p. 4; March 1995, p. 4.

52. "Direct Marketing Flow Chart," *Direct Marketing*, January 1986, p. 17; December 1991, p. 2; March 1995, p. 4.

53. "Direct marketing," *Advertising Age*, July 19, 1982, p. M-27.

54. Robert Delay, "Direct Marketing—Way Beyond Catalogs," *Advertising Age*. April 30, 1980, p. 188. In a 1990 book, Peter Winter reflected the new high-tech aspect of

direct marketing with a definition that updated the 1982 version noted earlier. It described direct marketing as "the distribution of goods, services and information to targeted consumers through response advertising while keeping track of sales, interest and needs in a computer database." Winter is quoted on p. 46 of Rapp and Collins, *The Great Marketing Turnaround*.

55. "An Icon at Nearly 100," *Industrial Engineer*, July 1, 2004, p. 16; F. John Reh, "Pareto's Principle—The 80-20 Rule."

56. Even accountants find the notion relevant to describe their bad clients. According to the accountancy author Ron Baker, the rule correctly predicts that "most accountants end up with a core group of frustrating clients who cause 80% of their problems." See "When Bad Clients Drive Out the Good," *Chartered Accountants Journal of New Zealand*, September 2003, p. 31.

57. Don Peppers and Martha Rogers, *The One to One Future* (Doubleday, 1993, pp. v–vi.

58. Jennifer Lawrence, "Continental Tries Airline Rebate Offer," *Advertising Age*, September 19, 1988, p. 3.

59. Laurie Freeman, "Direct Contact Key to Building Brands," *Advertising Age*, October 25, 1993, p. S-2.

60. Ibid.

61. "Using Databases to Seek Out the (Brand) Loyal Shoppers," *Promo*, February 1, 1995.

62. Freeman, "Direct Contact Key to Building Brands."

63. Alice Cuneo, "Frequent-Buyer Programs Are Still Flying High," *Advertising Age*, March 20, 1995.

64. Kelly Shermach, "Large and Small Retailers See Value in Data-Base Marketing," *Data-Base Marketing*, September 25, 1995, p. 8.

65. Freeman, "Direct Contact Key to Building Brands."

66. "The Appeal of Direct Response" (editorial), *Advertising Age*, October 16, 1995, p. 14.

67. Ibid.

68. Ibid.

Chapter 4

1. Valerie Mackie, "Electronic Newsstand to Add Ads," *Advertising Age*, November 1, 1993, p. 22. On the libertarian nature of early internet culture, see Frances Bula,

"Two Sides to Sex Link," *Vancouver Sun*, January 30, 1993. See also Esther Dyson, *Release 2.0* (Broadway Books, 1997).

2. Aaron Zitner, "Internet Losses: The Traditional Rules Don't Apply in Running On-Line Enterprises," *Boston Globe*, November 22, 1994.

3. Ibid. The most famous usenet flaming incident took place in 1994, when Laurence Canter and Martha Siegel, husband-and-wife lawyers from Arizona, became the targets of written online attacks after they posted advertisements on internet bulletin boards offering to help immigrants get green cards.

4. Mackie, "Electronic Newsstand to Add Ads."

5. Frank Beacham, "Marketing on the Internet a Daunting Prospect," *Advertising Age*, December 13, 1993; telephone interview with Dale Dougherty, November 30, 2004. The World Wide Web—an area of the internet designed to allow users to reach and share documents through the use of links—was created in the late 1980s. Exciting as that was, it was the Mosaic browser, invented at the University of Illinois' National Center for Supercomputing Applications, that caused a flood of public enthusiasm. Mosaic, introduced in the spring of 1993, and Netscape's commercial version, which soon followed, allowed web users to see both text and graphics instantly, and the click-and-see navigation approach made accessing materials much easier than it had been in the internet's text-based regime.

6. See, for example, E. S. Cohen, "Viewpoint/Letters: New Media Mirages," *Advertising Age*, November 1, 1993, p. 28.

7. Martin Nisenholtz, "Letters: Interactive Is Very Active," *Advertising Age*, November 15, 1993, p. 26.

8. Edwin Artzt, "The Future Of Advertising," *Vital Speeches*, September 1, 1994 (delivered May 12, 1994).

9. Ibid.

10. Donald Libby, "Cruising and Vacuuming the Internet," *internet Marketing News*, January 20, 1995, p. 7.

11. Patricia Riedman, "Interactive: Juno Sues over Falsified Addresses," *Advertising Age*, December 1, 1997, p. 68.

12. See John Schwartz, "Giving Web a Memory Cost Its Users Privacy," *New York Times*, September 4, 2001; Roger Clark, "Cookies," at http://www.anu.edu; Greg Elmer, *Profiling Machines* (MIT Press, 2004), pp. 118–119.

13. Matt Carmichael, "Interactive," *Advertising Age*, November 18, 1996, p. 50.

14. See Greg Elmer, *Profiling Machines* (MIT Press, 2004), pp. 118–119.

15. James Staten, "Navigator Tricks Raise Concerns," *MacWeek*, March 18, 1996, p. 18.

16. Graeme Browning, "A Tangled Web," *National Journal*, September 7, 1996, p. 1899.

17. Ibid.

18. "Year in Review: Interactive," *Advertising Age*, December 22, 1997, p. 21.

19. Ibid.

20. Thom Weidlich, "DMers and Net Heads Try to Warm Up to One Another," *Direct*, November 1, 1999.

21. "Year in Review: Interactive," *Advertising Age*, December 22, 1997, p. 21.

22. The new infrastructure also included specialized internet advertising agencies that were guiding the movement of companies online. NetRatings, Media Metrix, @plan, and Relevant Knowledge, which vied to become the Nielsen of the web, claimed to note what websites a random sample of online Americans visited on the web. ABC Interactive, BPA Interactive, and I/PRO worked alone and together to verify the claims by individual websites of their numbers of visitors. Matchlogic was involved in verifying whether the ads that companies said they were serving did indeed go out to web users.

23. Quoted in Stephanie Thompson, "Cereal Makers Entice Online Kids," *Advertising Age*, July 3, 2000, p. 20. See also Jack Neff, "Superstar Lesley Mansford and Nick Rush: Pogo Keeps Loyalists Coming Back for More," *Advertising Age*, March 26, 2001, p. S14; Tobi Elkin, "The Fame Game," *Advertising Age*, June 25, 2001, p. 36.

24. See Renee Dye, "The Buzz on Buzz," *Harvard Business Review*, November-December 2000, p. 139ff.

25. Source of quotation: www.deiworldwide.com, January 21, 2005.

26. See Jack O'Dwyer, "Word of Mouth PR Crosses Ethical Lines," *O'Dwyer's PR Services Report*, January 2005, p. 1.

27. "Year in Review: Interactive," *Advertising Age*, December 22, 1997, p. 21.

28. Don Peppers and Martha Rogers, *The One to One Future* (Doubleday, 1993), p. 259.

29. Weidlich, "DMers and Net Heads Try to Warm Up to One Another."

30. Beth Negus, "A DM E-volution," *Direct*, November 1999.

31. Ibid.

32. Ibid.

33. Stephen Barrington, "Canadian Marketers Take Steps to Foster Web Commerce," *Advertising Age*, November 10, 1997, p. S-35.

34. Weidlich, "DMers and Net Heads Try to Warm Up to One Another."

35. Ira Teinowitz "The Pressure Is Building in Washington to Limit the Kind of Personal Information Marketers Can Ask of Kids Online," *Advertising Age*, June 23, 1997, p. 16.

36. Ira Teinowitz, "FTC Applies More Pressure to Sites Marketing to Kids: Government Demands Tougher Guidelines for Online Privacy," *Advertising Age*, June 23, 1997, p. 23.

37. Ira Teinowitz, "FTC Chief Asks Congress to Ensure Privacy on Web," *Advertising Age*, June 8, 1998, p. 53.

38. Jennifer Gilbert and Ira Teinowitz, "Privacy Debate Continues to Rage; Doubleclick Taking Heat for All Ad Networks," *Advertising Age*, February 7, 2000, p. 44.

39. Ira Teinowitz, "Consumers to Be Notified about Profiling," *Advertising Age*, November 15, 1999, p. 52.

40. Jennifer Gilbert and Ira Teinowitz, "Marketers Address Web Concerns," *Advertising Age*, February 21, 2000, p. 1.

41. John Kamp, quoted in Gilbert and Teinowitz, "Marketers Address Web Concerns."

42. Gilbert and Teinowitz, "Marketers Address Web Concerns."

43. This discussion is based on my reading of hundreds of web privacy policies and on my systematic analysis of sites with substantial child audiences. See Joseph Turow, Privacy Policies on Children's Websites: Do They Play by the Rules? (Annenberg Public Policy Center, 2001).

44. See Joseph Turow, Americans and Online Privacy: The System Is Broken (Annenberg Public Policy Center, 2003).

45. Joseph Turow, The Internet and the Family: The View from Parents, the View from the Press (Annenberg Public Policy Center, 1999).

46. Joseph Turow and Lilach Nir, The Internet and the Family 2000: The View from Parents, the View from Kids (Annenberg Public Policy Center, 2000).

47. Ibid.

48. Alan F. Westin, "Social and Political Dimensions of Privacy," *Journal of Social Issues* 59 (2003), no. 2: 431–453.

49. Beth Negus Viveiros, "On the Bright Side," *Direct*, October 15, 2004, p. 14.

50. See, for example, Bart A. Lazar, "Can Spam Creates Quandary, Opportunities," *Marketing News*, June 15, 2004, p. 6.

51. See, for example, Lazar, "Can Spam Creates Quandary, Opportunities."

52. Louis Mastria, quoted in D.L.V., "Regulatory, Legislative, Economic," *Marketing News*, January 15, 2004, p. 19.

53. Quoted in Anthony Bianco, "The Vanishing Mass Market," *Business Week*, July 12, 2004, p. 60.

54. James Stengel, "Stengel's Call to Arms," *Advertising Age*, February 16, 2004, p. 16; Jack Neff and Lisa Sanders, "It's Broken," *Advertising Age*, February 16, 2004, p. 1.

55. Weidlich, "DMers and Net Heads Try to Warm Up to One Another."

56. The state of Utah, in fact, made adware illegal under a "Spyware Control Act" that specifically banned adware as spyware. The 2004 statute banned outright the use of a "context based triggering mechanism to display an advertisement that partially or wholly covers or obscures paid advertising on an internet Web site in a way that interferes with a user's ability to view the internet Web site." The ban would apply even if a user explicitly agreed to the download and the program could be easily removed. It was a law that startled mainstream advertisers and software providers, who could see the utility of sending ads that way in an opt-in manner. In 2005 the law being challenged before the Utah Supreme Court by WhenU.com on numerous federal and state statutory and constitutional grounds. Technology attorney Alan Friel suggests that the upholding of the Utah law would spark marketers to encourage federal legislation that distinguished between spyware and adware and allowed the triggering of ads as long as they provided notice to consumers and some version of choice (source: interview with Alan Friel). See also Alan Friel, "Privacy Patchwork," Marketing Management, November-December 2004, p. 48.

57. Brian Deagon, "Sly "Spyware' Sneaks Past Viruses as No. 1 Complaint of PC Users," *Investor's Business Daily*, July 1, 2004.

58. Ibid.

59. For a justification of cookie activities by these firms, see www.networkadvertis ing.org/optout_nonppii.asp.

60. "Tacoda Systems Launches Next Generation Audience Management System," *PR Newswire*, November 4, 2003.

61. Tremor website, accessed through Wayback Machine (http://web.archive.org/web/*/http://tremor.com).

62. Ross Fadner, "inGamePartners Expands Advergaming Reach with New Deals," Media Daily News, September 13, 2004.

63. John Fedderman, "Banners Aren't Dead," *iMedia Connection*, September 28, 2004.

64. "Case Study: American Airlines" (http://epiphany.com/products/downloads/Amer_Airlines_CS.pdf), p. 3.

65. Ibid., p. 4.

66. Ibid., p. 4.

67. Ibid., p. 5.

68. Lorraine Calvacca, "Data on Demand," *Direct*, July 1, 2004, p. 9.

69. Robert Buderi, "E-Commerce Gets Smarter," *Technology Review*, April 2005.

70. "Customer Recognition Solutions," http://acxiom.com, January 5, 2005.

71. Calvacca, "Data on Demand," p. 9.

72. "Customer and Information Management," Acxiom website, January 5, 2005.

73. "Prospect Marketing Solutions," http://acxiom.com, January 5, 2005.

74. John Ragsdale, "From Satisfaction to Loyalty," *Best Practices* (Forrester Research), December 27, 2004, p. 4.

75. Jack Neff, "P&G Extends Online Custom Publishing," *Advertising Age*, March 22, 2004, p. 24.

76. Ragsdale, "From Satisfaction to Loyalty," p. 4.

77. Ibid.

78. Ibid.

79. Calvacca, "Data on Demand," p. 9.

80. David Vise, "What Lurks in Its Soul?" *Washington Post*, November 13, 2005.

81. FaceBook.com privacy policy, accessed December 20, 2005.

82. Telephone interview with Kurt Viebrans, January 5, 2005.

Chapter 5

1. Scott Donaton, *Madison & Vine* (McGraw-Hill, 2004), pp. 181–182.

2. Scott Donaton, "CBS, TV's New Bauble Seller, Boasts When It Should Blush," *Advertising Age*, June 28, 1999, p. 32.

3. Randall Rothenberg, "Technologies to Provide Win-Win Fare for Programmers, Consumers," *Advertising Age*, July 19, 1999, p. 15.

4. Donaton, *Madison & Vine*, p. 3.

5. Mike Shaw, head of ABC-TV advertising, at Media Summit conference, February 9, 2005.

6. Kimber Sterling, quoted in Anne M. Mack, "Interactive Quarterly," *AdWeek*, September 20, 2004.

7. Anne Moncreiff Arrante, "Ratings Aside, U.S. TV Takes Hold," *Advertising Age*, October 30, 1989, p. 30.

8. Josh Bernoff, "Digital Recorders Take Flight," Forrester Market Overview, April 16, 2004, p. 4.

9. Anne M. Mack, "Interactive Quarterly," *AdWeek*, September 20, 2004.

10. T. L. Stanley, "The Buzz," *Advertising Age*, June 7, 2004, p. 53.

11. See Claire Atkinson, "Jeff Lukas," *Advertising Age*, September 27, 2004, p. S-28; "IAG's Top 10 Most-Recalled Product Placements in Network Sitcoms September 20–November 29, 2004," *Advertising Age's Madison+Vine*, December 12, 2004; "IAG's Top 10 Most-Recalled Product Placements in Network Dramas, September 20–November 29, 2004," *Advertising Age's Madison+Vine*, November 17, 2004.

12. Scott Donaton, *Madison & Vine* (McGraw-Hill, 2004), pp. 33–34.

13. James Curtis, "Brands Are Now Looking Beyond Traditional TV Ad and Sponsorship Work to Connect with Audiences," *Marketing*, September 11, 2003, p. 20.

14. Tony Gnoffo, "Technology Forces Television Advertisers to Re-evaluate Methods," *Philadelphia Inquirer*, January 23, 2005. See also Kate Fitzgerald. "Growing Pains for Placements," *Advertising Age*, February 3, 2003, p. S2.

15. Joel Brinkley, "Do Viewers Even Want to Interact with TV?" *New York Times*, February 7, 2000.

16. Wayne Walley, "Big Three Poll with 900 numbers," *Electronic Media*, March 17, 1991, p. 30.

17. Martin Renzhofer, "'Looney Tunes' and 'Brady Bunch' Offer Security in Economically Insecure World," *Salt Lake Tribune*, March 29, 2001.

18. "Viewers Spare *Law & Order* Character," *Tulsa World*, October 27, 2004.

19. Walley, "Big Three Poll with 900 numbers"; "Viewers Spare *Law & Order* Character."

20. Susan Young, "Your Turn to Pick Survivor Millionaire," *Alameda Times-Star*, May 11, 2004.

21. Richard L. Eldredge, "Lovable 'Survivor' Hugely Popular, Too," *Atlanta Journal-Constitution*, May 21, 2004.

22. DBS: direct broadcast satellite.

23. Martha Bennett, "Interactive Television: Moving Forward, at a Snail's Pace," Forrester IdeaByte, November 1, 2002.

24. For other discussions of Bernoff's report, see Glen Dickson, "Forrester Foresees 'Smart TV,'" *Broadcasting & Cable*, July 17, 2000, p. 10; Andrew Humphreys, "Direct to Cyberconsumer," *Med Ad News*, November 1, 2000, p. 1.

25. Joshua Bernoff, "Smarter Television," Forrester Report, July 2000, p. 10.

26. Ibid., p. 13.

27. Ibid., p. 13.

28. Ibid., p. 16.

29. Interview with Pat Ruta, vice president for business development at Visible World, February 11, 2005.

30. Brodie Keast, quoted in Tobi Elkin, "Madison + Vine: Getting Viewers to Opt In, Not Tune Out," *Advertising Age*, November 4, 2002, p. 10.

31. Sarah McBride, "Games for the Couch Potato," *Wall Street Journal*, January 17, 2005.

32. Ibid.

33. Ibid.

34. Dan Campbell of MTV and Marcia Zellers of AFI, quoted in ibid.

35. Anne M. Mack, "Interactive Quarterly," *AdWeek*, September 20, 2004.

36. Jason Kuperman, quoted in ibid.

37. Mack, "Interactive Quarterly."

38. Tony Gnoffo, "Technology Forces Television Advertisers to Re-Evaluate Methods," *Philadelphia Inquirer*, January 23, 2005.

39. Comcast Spotlight, "Split the Message, Increase Your Impact," February 13, 2005.

40. Conversation with Dana Runnells, Comcast Spotlight, February 11, 2005.

41. Jean Halliday, "Ford Dealers Test Custom Cable Ads," *Advertising Age*, October 18, 2004, p. 6.

42. Jack Neff, "Addressable TV Meets with Agency and Marketer Resistance," *Advertising Age*, March 15, 2004, p. 12.

43. Interview with Pat Ruta, February 11, 2005.

44. Interview with Gerrit Niemeijer, February 14, 2005.

45. Ibid.

46. See Cable TV Privacy Act of 1984, Section 551, "Protection of Subscriber Privacy," part c.

47. Ibid.

48. Comcast Spotlight, "Split the Message, Increase Your Impact," February 13, 2005.

49. Time Warner Cable privacy policy.

50. Conversation with Hank Oster of Comcast Spotlight, February 28, 2005.

51. Interview with Gerrit Niemeijer. February 14, 2005.

52. Dan Nova, quoted in Janet Whitman, "New Ad Technology Tailors TV Ads to Specific Viewers," *Marketing News*, November 15, 2004, p. 35.

53. Eric Schmitt, quoted in Whitman, "New Ad Technology Tailors TV Ads to Specific Viewers."

54. Richard Linnett, "AdAges," *Advertising Age*, March 4, 2004, p. 36.

55. See Wayne Friedman, "Eagle-Eye Marketers Find Right Spot, Right Time," *Advertising Age*, January 22, 2001, p. S-2.

56. Rich Thomaselli, "Targeting New Revenue: Sports Tech Firms Seek Ad Sponsors," *Advertising Age*, May 13, 2002, p. 6; Princeton Video Imaging website (www.pvimage.com).

57. Interview with Pat Ruta.

58. Tracy Swedlow, "Verizon to Use Microsoft TV for Its 'FiOS' TV Service," *Tracy Swedlow's itvt newsletter*, February 18, 2005.

59. Ibid.

60. Ibid.

61. Gnoffo, "Technology forces television advertisers to re-evaluate methods."

62. Bob Garfield, "Garfield's Bobbys," *Advertising Age*, January 3, 2005, p. 25.

63. Mack, "Interactive Quarterly," *AdWeek*, September 20, 2004.

64. Remarks at Advertising Strategies in the Diversified Digital Culture, session A, McGraw-Hill Media Summit, New York, February 9, 2005.

65. Ibid.

66. This discussion draws heavily on an interview with Dena Kaplan published in *Tracy Swedlow's itvt newsletter*, Part II, February 18, 2005.

67. Mack, "Interactive Quarterly."

68. Erik Davis, vice president of marketing at Extend Media, interviewed in *Tracy Swedlow's itvt newsletter*, part II, February 18, 2005.

Chapter 6

1. See Gary McWIlliams, "Minding the Store: Analyzing Customers," *Wall Street Journal*, November 8, 2004. See also "Retail Brief: Best Buy Co.: Net Increases 49% amid Lift From Customer Centric Stores," *Wall Street Journal*, June 15, 2005.

2. "Dorothy Lane Markets Gives a Personal Touch to Loyalty Members," Progressivegrocer.com, October 21, 2004.

3. Richard Shulman, "Picking Your MVPs," *Progressive Grocer*, March 1, 2005.

4. Beth Negus Viveiros, "Invited Guests," *Direct*, December 1, 2004, p. 16.

5. Letitia Baldrige, "I Shopped Them All," *New York Times*, March 8, 2005.

6. John Reese, "Wal-Mart Still a Winner," The Street.com, March 8, 2005; Cecil Johnson, "Huge Population, Growing Economic Clout Making China Force in Markets," *Fort Worth Star-Telegram*, March 3, 2005.

7. Reese, "Wal-Mart Still a Winner."

8. Ibid.

9. "Supercenters' Success Ups Ante," *Chain Store Review*, April 26, 2004, p. 64.

10. Burt Flickinger, quoted in "Chapter 11 Shoe Drops on Winn-Dixie," ProgressiveGrocer.com, February 23, 2005.

11. "Supercenters' Success Ups Ante."

12. Conversation with retail consultant Richard Shulman, March 8, 2005.

13. Constance L. Hays, "What They Know About You," *New York Times*, November 14, 2004.

14. In *China Inc.* (Scribner, 2005), Ted Fishman writes that Chinese factories are by far the most important and fastest-growing sources for Wal-Mart; in 2003 alone the company purchased $15 billion in goods from Chinese suppliers. Fishman says Wal-Mart has 560 buyers in China, and that 10–13 percent of China's exports to the United States wind up at Wal-Mart. He cites an article in the *Washington Post* as asserting that "80 percent of the 6,000 factories in Wal-Mart's worldwide database of suppliers are in China." He adds that "Wal-Mart's growth as an economic force is inseparable from China's rise as a manufacturing giant."

15. "Wal-Mart Bets on RFID as the Wave of the Future," *Chain Drug Review*, July 19, 2004.

16. Richard Shulman, "Budgeting for a Moving Target," *Progressive Grocer*, October 1, 2004.

17. Hays, "What They Know About You."

18. Conversation with Gib Carey, March 16, 2004.

19. Conversation with Edward Fox of J. C. Penney Center for Retail Excellence and Southern Methodist University, March 14, 2005.

20. Ibid.

21. Quoted in Charles Fishman, "The Wal-Mart You Don't Know," *Fast Company*, December 1, 2003, via Factiva. See also Dan Mitchell, "Manufacturers Try to Thrive on the Wal-Mart Workout," *New York Times*, February 20, 2005.

22. Segmentation Analysis: "P$cycle," Claritas website, accessed March 23, 2005.

23. Rajkumar Venkatesan and V. Kumar, "A Customer Lifetime Value Framework for Customer Selection and Resource Allocation Strategy," *Journal of Marketing*, fall 2004: 106–125.

24. Michael Johnson and Fred Selnes, "Customer Portfolio Management," *Journal of Marketing*, spring 2004: 1–17.

25. Venkatesan and Kumar, "A Customer Lifetime Value Framework."

26. Conversation with Matthew Gracie of Riggs Bank, March 21, 2005.

27. Ibid.

28. Segmentation Analysis: "P$cycle."

29. Conversation with Community Banking Manager David Kearsley and Regional President David Ehst of Sovereign Bank, March 21, 2005.

30. Segmentation Analysis: "P$cycle."

31. "Data Products: Income Producing Assets," Claritas website, accessed March 23, 2005.

32. Segmentation Analysis: "P$cycle,"

33. Conversation with Kearsley.

34. Segmentation Analysis: "P$cycle."

35. Charles Wendell, "Facing Channel Management Challenges," *American Banker*, November 17, 2004, p. 23a.

36. Charles Wendell, "How to Move Customers Out of Branches," *American Banker*, April 13, 2004, p. 15A.

37. Bill Stoneman, "Marrying Segmentation, Channel Management," *American Banker*, November 17, 2004, p. 3a.

38. Wendell, "Facing Channel Management Challenges."

39. "Green Checking," at www.citizensbank.com (accessed March 24, 2005).

40. "Citizens Circle Checking," at www.citizensbank.com (accessed March 24, 2005).

41. "Citizens Circle Gold Checking with High Interest," at www.citizensbank.com (accessed March 24, 2005).

42. Wendell, "Facing Channel Management Challenges."

43. Conversation with Laura Gainey. See also "Size Really Does Matter When You're Calling Some of Canada's Largest Financial Institutions," CP, July 26, 2004.

44. Stoneman, "Marrying Segmentation, Channel Management," p. 3a.

45. Interview with Kevin Perkis, Royal Bank of Canada, March 22, 2005.

46. Lavonne Kuykendall, "Repackaging to Make Clients More Profitable," *American Banker*, June 6, 2002, p. 2.

47. Mark N. Hendrix, "Segmenting, Comfy Chairs and Other Ways to Improve Interaction," *American Banker*, January 23, 2004, p. 7.

48. Conversation with Frank Berman, head of marketing, Bloomingdale's, March 25, 2005.

49. Richard H. Levey, "Bloomingdale's Goes for the Best," *Direct*, January 1, 2004, p. 1.

50. Ibid.

51. Ibid.

52. Conversation with Karl Bjornson, retail consultant and Senior Manager, Kurt Salmon Associates, March 21, 2005.

53. Conversation with Karen Meletam, Manager of Media Relations, Wakefern Food Corporation, March 30, 2005.

54. Conversation with Michelle Bauer of Catalina Marketing, March 2005.

55. Bob Francis, "After the Sale," *Brandweek*, March 31, 2003.

56. Ibid.

57. William Atkinson, "Tagged," *Risk Management* 7:51 (July 1, 2004), p. 12.

58. Ibid.

59. Howard Wolinsky, "P&G, Wal-Mart Store Did Secret Test of RFID," *Chicago Sun Times*, November 9, 2003.

60. Ibid.

61. Atkinson, "Tagged."

62. Scott Klein, CEO of Information Resources Inc., quoted in Jack Neff, "Why Some Marketers Turn Away Customers," *Advertising Age*, February 14, 2005, p. 1.

63. Larry Aronson of Cartwheel, quoted in in Jack Neff, "Why Some Marketers Turn Away Customers," *Advertising Age*, February 14, 2005, p. 1

64. Richard Shulman, "Picking Your MVPs," *Progressive Grocer*, March 1, 2005.

65. Conversation with Michelle Bauer, March 2005

66. Shulman, "Picking Your MVPs."

67. Interviews with Pedro, a clerk at a Dallas Albertson's store (March 10, 2005), and with Ray Nichols, director of print advertising for Albertson's in Dallas (March 11, 2005).

68. Faith Weiner, senior director of government and media relations for Stop & Shop, quoted in "Stop & Shop to Roll Out 'Shopping Buddy' to 20 Stores in 1Q 05," Progressive Grocer online edition, October 14, 2004.

69. Ibid.

70. Conversation with Bjornson.

Chapter 7

1. Bob Garfield, "The Chaos Scenario," *Advertising Age*, April 4, 2005, p. 1.

2. Richard Fielding, quoted in Jon Gertner, "Our Ratings, Ourselves," *New York Times Magazine*, April 10, 2005.

3. https://www.donotcall.gov/FAQ/FAQDefault.aspx, accessed April 21, 2005.

4. http://www.ftc.gov/ftc/consumer.htm, accessed April 21, 2005.

5. http://www.ftc.gov/privacy/index.html, accessed April 21, 2005.

6. "Gramm-Leach-Bliley Act's Safeguards Rule against Mortgage Companies," http://www.ftc.gov/opa/2004/11/ns.htm, accessed April 21, 2005.

7. http://www.infowars.com/cashless_society.htm, accessed April 25, 2005.

8. http://www.nocards.org/welcome/index.shtml, accessed April 22, 2005.

9. http://www.commercialalert.org/index.php/article_id/AboutUs, accessed April 22, 2005.

10. http://www.commercialalert.org/blog/, accessed April 22, 2005.

11. http://www.commercialalert.org/index.php/category_id/1/subcategory_id/53/article_id/98, accessed April 22, 2005.

12. http://www.commercialalert.org/aboutus.pdf See also John Schwartz, "Offer of Free Computers for Schools Is Withdrawn," *New York Times*, November 2, 2000.

13. http://www.junkbusters.com/summary.html, accessed April 25, 2005.

14. "EPIC Opens West Coast Office," EPIC press release, March 7, 2005.

15. "ChoicePoint Vows to Tighten Controls," *New York Times*, February 22, 2005.

16. Jon Swartz, "Rules Aimed at Digital Misdeeds Lack Bite," *USA Today*, April 11, 2005.

17. The FTC and "marketing" appeared in the *Journal of Economics & Management Strategy* and in the *Journal of Organizational and End User Computing*.

18. "Make It Count: A Consumer Guide," *Seattle Times*, November 8, 2004.

19. Carolyn Bigda, "Act Fast to Prevent, Limit Damage from Identity Theft," *Chicago Tribune*, November 7, 2004.

20. Ariana Eunjung Cha, "Finding Fewer Happy Returns, Retailer Databases Limit Some Shoppers," *Washington Post*, November 7, 2004.

21. Jolayne Houtz, "List Brokers Selling Children's Personal Information," *Chattanooga Times Free Press*, September 19, 2004.

22. Paul Meyer and Matt Stiles, "Has Your Identity Been Sold?" Dallas Morning News, September 26, 2004. See also Matt Richtel, "Credit Card Theft Is Thriving Online as Global Market," *New York Times*, March 13, 2002.

23. See, for example, Elaine Walker, "Shoppers' Data Stolen from DSW Shoe Stores," *Miami Herald*, April 20, 2005.

24. See Joseph Turow, Lauren Feldman, and Kimberly Meltzer, Open to Exploitation: American Shoppers Online and Offline (Annenberg Public Policy Center, 2005); Joseph Turow, Americans and Online Privacy: The System Is Broken (Annenberg Public Policy Center, 2003); Joseph Turow and Michael Hennessy, "Internet Privacy and Institutional Trust: Insights From a National Survey," *New Media & Society*, in press.

25. Robert Buderi, "E-Commerce Gets Smarter," *Technology Review*, April 2005, p. 4.

26. IBM, "Hippocratic Database," at http://www.zurich.ibm.com (accessed December 30, 2005).

27. Buderi, "E-Commerce Gets Smarter."

28. Ibid.

29. Ann N. Mack, "How Online Retailers Are Luring Holiday Shoppers," *Adweek*, December 6, 2004.

30. Bob Greenberg, *Adweek*, January 3, 2005.

31. "Search Isn't the Death of Advertising, It's the Future," *Marketing Week*, January 6, 2005, p. 24.

32. David Robinson, "Shop Till You Drop Your Mouse," *Buffalo News*, March 14, 2005.

33. LeeAnn Prescott, "Shopping Around On Line," *Chain Store Age*, March 2005, p. 78.

34. Ibid.

35. Alan Rimm-Kaufman, quoted in Heather Retzlaff, "Shopping Round," *Catalog Age*, January 2005, p. 26.

36. Heather Retzlaff, "Shopping Around," *Catalog Age*, January 2005, p. 26.

37. Angie McLoskey, quoted in Retzlaff, "Shopping Around."

38. "Kelkoo: Brand Not Price Is Top Priority for Online Retail," *Marketing Week*, November 11, 2004, p. 17.

39. Lauren Freedman, "Multichannel Goes Mainstream," *Catalog Age*, May 1, 2004, p. 5.

40. Karen Kroll, "The Price of Discounting," *Catalog Age*, March 1, 2005, p. 1.

41. W. Baker, W., M. Marn, M., and C. Zawada, "Price Smarter on the Net," *Harvard Business Review* 79 (2001), no. 2: 122–127.

42. M. Kung, M., K.B. Monroe, and J. L. Cox. "Pricing on the Internet," *Journal of Product and Brand Management* 11 (2002), no. 4/5: 274–287.

43. See J. Adamy, "E-tailer Price Tailoring May Be Wave of Future," *Chicago Tribune*, September 25, 2000.

44. William McGee, "Major Travel Sites Face Credibility Crunch," www.consumerwebwatch.org.

45. Personal communication to the author, May 2, 2005.

46. "Kelkoo: Brand Not Price Is Top Priority for Online Retail," *Marketing Week*, November 11,2004, p. 17.

47. Retzlaff, "Shopping Around."

48. "Kelkoo: Brand Not Price Is Top Priority for Online Retail."

49. Retzlaff, "Shopping Around."

50. Howard Baker, quoted in Shankar Gupta, "Consumer-Generated Blogger Sidelines E-Commerce Site," *OnlineMediaDaily*, December 2, 2005.

51. Ibid.

52. Shankar Gupta, "Reviews by Online Shoppers Spark Cynicism," *OnlineMediaDaily*, December 8, 2005.

53. Ibid.

54. Matt Rand, "Comparison Shopping on Sale," Forbes.com.

55. Robinson, "Shop Till You Drop Your Mouse."

56. Jeffrey Grau, "Online Consumer Selling," *eMarketer*, February 2005, p. 6. Forrester Research says the figure is $137 billion, but its percentage increase is the same: 20%. See Carrie Johnson, "2004 U.S. eCommerce," *Forrester Trends*, January 27, 2005, p. 1.

57. Grau, "Online Consumer Selling."

58. Carrie Johnson, "Multichannel Retailing Best Practices," *Forrester Best Practices*, September 15, 2005, p. 2.

59. Mike Gilpin, "The Interaction Platform," *Forrester Trends*, October 4, 2004, p. 2.

Chapter 8

1. Email to the author from a *Washington Post* reader, June 19, 2005.

2. Shankar Gupta, "Microsoft Usher in Ad-Supported Software," *Mediapost Daily News*, November 3, 2005.

3. See, for example, Oscar Gandy Jr., "The Real Digital Divide: Citizens vs. Consumers," in *The Handbook of New Media*, ed. L. Lievrow and S. Livingstone (Sage, 2002); Oscar Gandy, "Data Mining, Privacy and the Future of the Public Sphere," Second Dixons Lecture, London School of Economics, November 7, 2002, available at www.lse.ac.uk.

4. Paul Gamble, Merlin Stone, and Neil Woodcock, *Up Close and Personal?* (Kogan Page, 1999), p. 296.

5. Turow et al., Open to Exploitation.

6. William Atkinson, "Tagged," *Risk Management* 7 (2004), no. 51: 12.

7. "Jump$tart Sees Surge in Legislation Promoting Financial Literacy" (Jump$tart press release, March 5, 2004).

8. See Educational Testing Service, "The No Child Left Behind Act: A Special Report," June 2002, available at http://ftp.ets.org; "Today's Events in Washington," *The Frontrunner*, July 21, 2004.

Index